FIBONACCI
ANALYSIS

Related titles also available from BLOOMBERG PRESS

Bloomberg MARKET ESSENTIALS
TECHNICAL ANALYSIS

FIBONACCI ANALYSIS

Constance BROWN

BLOOMBERG PRESS

NEW YORK

This publication contains the author's opinions and is designed to provide accurate and authoritative information. It is sold with the understanding that the author, publisher, and Bloomberg L.P. are not engaged in rendering legal, accounting, investment-planning, or other professional advice. The reader should seek the services of a qualified professional for such advice; the author, publisher, and Bloomberg L.P. cannot be held responsible for any loss incurred as a result of specific investments or planning decisions made by the reader.

First edition published 2008

3 5 7 9 10 8 6 4

Library of Congress Cataloging-in-Publication Data

Brown, Constance M.
Fibonacci analysis/Constance Brown.
    p. cm.—(Bloomberg market essentials)
  Includes bibliographical references and index.
  Summary: "Fibonacci Analysis gives traders a first step toward applying Fibonacci analysis in the market. The book covers key elements of this technical analysis tool. Traders will get up to speed quickly on its unique benefits. The major ratios between numbers in the Fibonacci sequence describe predictable market patterns. Fibonacci analysis analyzes these movements to identify future market prices"—Provided by publisher.
  ISBN 978-1-57660-261-4 (alk. paper)
  1. Investment analysis. 2. Speculation. 3. Fibonacci numbers. I. Title.

HG4529.B762 2008
332.63'2042—dc22

2008021539

Geometry has two great treasures: one is the theorem of Pythagoras; the other, the division of a line into extreme and mean ratio. The first we may compare to a measure of gold; the second we may name a precious jewel.

— Johannes Kepler (1571–1630)

Contents

List of Illustrations

Acknowledgments

FIBONACCI ANALYSIS IS a personal passion of mine, and I would like to thank the entire team at Bloomberg Press—including David George, JoAnne Kanaval, Dru-Ann Chuchran, and Leah Shriro—for raising the quality bar to the highest standard possible. My special thanks go to Stephen Isaacs. With each book we've done together, my trust in him grows.

Another important part of the team behind this book was Market Analyst Software, the Australian software vendor that has developed features and tools to push this subject past industry conventions. I would particularly like to thank Jonathan Garrett and Matthew Humphreys, who have struggled through many versions of the software development process to support my many requests. None of this would be possible without the support of the company's founder, Mathew Verdouw, who is committed to creating a revolution in technical analysis.

Last, but not least, is the personal support I was given behind the scenes from enthused colleagues and friends. That must also include the little Trade Street Gallery Coffeehouse in Tryon, NC, which attracts phenomenally accomplished Ivy League professors, physicists, business executives, and artisans and musicians who make an impact in such diverse fields as the Broadway theater and specialized fabrics protecting the world's firefighters. The Trade Street Gallery Coffeehouse is a mecca of talents, and the positive energy and enthusiasm is indeed infectious.

A lot of people have cared about and devoted their best efforts to this work. I hope this book will stand up to the test of time and become worn and dog-eared from years of use.

Introduction

Traders have been asking me for more than ten years to write this book. Being a fund manager, there was no chance I would write a comprehensive book on the price-projection methods I used or the research that was developing from the methodology. But times change and a wise partner of mine asked, "What is your business exit strategy going to be?" He explained that all the knowledge and wisdom you have gained over the years will be lost if you have no plan to pass it on to the next generation. It is not enough to come, extract from the markets, and then hide in a cave with the best gems you have polished over the years tucked safely away in the deepest partitions of your mind. What good was the journey if someone does not pick up the flag and continue to move it forward?

I never felt ready to put into print the most important work that the markets had trained me to use as the cornerstone of my various geometric methods. There was always more I needed to prepare before unveiling this work to my peers. It is alarming how time seems to fly by and indeed odd how someone becomes branded a specialist. One day, I was just working hard to catch the leaders in my craft, and then one morning, I awoke to find that the person looking back at me in the mirror was the one being hounded by others to reveal a deep hidden secret she held. There were no secrets except what was earned through hard work and a relentless passion to gain a better understanding of why something worked with each new market experience.

The final chapters of this book where completed at the end of December 2007. Book production schedules being what they are, this book could not be released until some months later. So with great interest I watched as 2008 began another significant equity decline, knowing this book contained the global market setup that helped to correctly identify the events that have since followed. Calling such moves before they happen and when is how my reputation grew within our industry. But how can one make such a claim well in advance? All the charts, targets, and methods within these pages where captured and described in real time to help you see why these methods never fail to astound and inspire me to continue to learn more.

Markets truly form a mathematical price grid when you know how to read the data. Every effort has been made to make these methods transparent for your use. But there will be much work needed on your part to master these techniques. Space constraints limited the amount of detail I could offer. Numerous examples cannot be developed when the body of work is so extensive. Therefore, other methods I use such as custom oscillators, market geometry, Gann analysis, and others are sparingly referenced to keep the precious few pages I was given focused on the primary subject. Markets from around the world have been selected, but please understand the methods do not vary if your specific market and time horizon was not discussed.

Make no misunderstandings; this book is likely the most important release I could offer you. If I were to write a blueprint of my trading style and the gems that have served me best through times of greatest market volatility, the work would be a trilogy. The first book would be the one in your hands today on Fibonacci techniques. The second book would be on harmonics. This book will leave you with an introduction to new concepts on harmonics, showing how we must push ourselves further to understand ratio analysis between market support and resistance zones on three axes. The third and final book would be about Gann analysis that incorporates the first two books, and then moves ahead to sophisticated methods of geometry and cycle analysis for timing precise pivot points and clarifying the price targets that have a common bond in major market moves.

For the reader who wants to continue this work forward, you will find the most extensive bibliography on this subject that our industry has seen to date. Every effort has been made to be transparent so those who want to learn more can do so. If the text references Plato's discussion of the Golden Section, the exact paragraph number in *Plato: Complete Works* has been identified. If a statement is made that the solar system has a Phi (or 1.618) relationship between all the planets, you will find a mathematical description of how this is true. Regarding Fibonacci, our industry has too often put the reader in the position that they must assume the author has found something to substantiate their claim. Not this time. There are no flippant generalities without proof, and readers will not be led to a dead end because the material offers no guidance on where the original thought or concepts were extracted. The methods are explained exactly as they have been applied for so many years. But that does not mean they will be easy. This book will require patience and work on your part. There are no shortcuts.

The hardest part for you is to toss out the old methods and assumptions you may be familiar with now so you can examine without prejudice the new. The more you can let go of old thinking, the easier you will find this book. However, be assured, many who know these methods from my semiprivate seminars will warn you that it takes time, effort, discipline, and practice. But your efforts and time will be rewarded, because these methods will answer the questions all traders need to ask. Where is the market going? At what level should my stop be entered? How much should be leveraged into the position based on the size of my trading account? What price level can tell me I am in trouble before my stop is hit? How much can or should I buy or sell given a second or third opportunity? Few traders can answer all these important questions. The accuracy of your answers will dictate if you will be around for long trading markets. The real goal for any trader is longevity. This race has no start or finish line. So choose wisely when and how you enter the race.

Some readers like to have a game plan before they step onto the field. You have eight chapters to read. Naturally, this book had to end on a Fibonacci number! The first chapter will provide you with the necessary background to build upon. Right up front, the experienced trader will see a chart application that the industry has overlooked. Experienced traders know one can add, subtract, multiply, and divide any Fibonacci number to obtain another, but the industry does not know how to apply this attribute.

Chapter 2 will teach you how to create a mathematical grid unique to a specific market that is not subjective. The concepts will help you understand why the conventional methods used in our industry today will only work some of the time. Market geometry must consider market expansion and contraction relationships within price and time axes. Methods in common use today severely handicap a trader. It is understandable why some people have abandoned their early efforts to understand and apply Fibonacci analysis. There are few books on the subject and none as comprehensive as the one before you. Joe DiNapoli first introduced me to the concept of confluence zones nearly eighteen years ago. Anyone who is familiar with his work will discover this book has moved far beyond his basic concepts. If you do not know what a confluence zone is at the moment, do not be concerned as no prerequisite reading is needed before you begin.

Chapter 3 will provide a detailed look at how to make price projections that you can have confidence in. The market will respect these targets, but they may not be the final objectives within larger market moves. That requires the added discussions and examples in Chapters 5 and 6.

Throughout the book is a tease that is eventually answered. Anyone who has crossed paths with this subject matter knows the Nautilus shell found in the Indian Ocean is one of the perfect examples in nature of the Fibonacci spiral. But just how do you put a spiral into a chart to make money? Why was this shell the primary symbol of the Pythagoreans, who were credited with the discovery of Phi (1.618)? Has anyone explained why the geometry of the Nautilus shell is so important to the study of sacred geometry? While we are on this trend of thought, why do the books on sacred geometry fail to describe where these shapes originate? So many questions may seem unanswered, but not after you have journeyed through these pages.

Your journey will take you to the Colosseum in Rome, ancient Baghdad and Babylonia, Cairo, the British Museum, and to hidden ancient papyrus of Egyptian high priests and kings. I have traveled the world to find answers, and the expenses were paid largely by the methods I share with you now. Then, as our journey begins to come to an end, I'll leave you at the opening of a new door for our industry. You will discover that the intervals between market support and resistance areas have harmonic properties. Price targets and time targets do not fall in a linear series. The important connecting intervals and market relationships have a geometry no one has put into print before now. It will force you to think outside the box. What will inspire you the most I cannot tell, but I know these pages offer the best this trader can pass on to those who are my peers, my colleagues, and the next generation of global movers and shakers who want to know more about Fibonacci analysis and how it helps us survive the turbulent markets of today and tomorrow. The blueprints are in your hands . . . what are you waiting for?

FIBONACCI
ANALYSIS

CHAPTER 1

The Mystery of Phi (1.618) and phi (0.618)

WHY IS THE GOLDEN RATIO (1.61803 . . .) of such interest to traders? Known by various names since the ancient Egyptians and Pythagoreans 570–490 BC, the first definition[1] comes from Euclid (325–265 BC); *The Divine Portion*, or *Divina Proportione*, by Luca Pacioli (1445–1517) was the earliest known treatise devoted to the subject and was illustrated by Leonardo da Vinci, who coined the name *section aurea* or the "golden section." By any of its historical names, it is of little interest to investors, but investors rarely face the day-to-day battlefield of rapid global market action. Traders on the frontlines, however, live and die on their ability to measure risk and control their capital drawdown. To enter the tough markets of today requires the foresight to call what price a market will move toward. Of equal importance is our need to know where a market should *not* be trading so we can execute a timely exit plan.

So how does an eloquent ratio that has captured people's interest from the time of antiquity help and influence the decisions a trader must face today? The Golden Ratio (**Figure 1.2**) is a universal law that explains how everything with a growth and decay cycle evolves. Be it the spiral of the solar system, the spiral of a hurricane (**Figure 1.1**), the growth pattern of a nautilus seashell, or the spiral that forms at the start of life as a fern begins to unfurl from the ground on a spring day. Plato (circa 427–347 BC) in the *Republic* asks the reader to "take a line and divide it unevenly." Under a Pythagorean oath of silence not to reveal the secrets of the mysteries, Plato posed questions in hopes of

1

FIGURE 1.1   **Hurricane Isabel**

Source: NASA

FIGURE 1.2   **The Golden Ratio**

provoking an insightful response. So why does he use a line, rather than numbers? Why does he ask you to divide it unevenly?

To answer Plato and how this may connect to the needs of all traders today, we first must understand ratio and proportion. All living things from large to small, *including markets*, must abide by the same divine blueprint. As a result, the more traders understand how a ratio called Phi (1.618) influences our lives, the clearer we can look into the face of a grander plan and witness a force acting upon our markets that demands our respect, comprehension, and humility. As a trader's knowledge deepens, he can see in advance a mathematical grid that determines future market movement. But few people truly understand the forces at play and never develop beyond the basic theory that is insufficient to adjust to the natural expansion and contraction cycles found within all markets in both price and time.

Ratio, Means, and Proportion

Ratio (*logos*) is the relation of one number to another, for instance 4 : 8, stated as "4 is to 8." However, proportion (*analogia*) is a repeating ratio that typically involves four terms, 4 : 8 :: 5 : 10, stated as "4 is to 8 is as 5 is to 10." The Pythagoreans called this a four-termed discontinuous proportion. The invariant ratio here is 1 : 2, repeated in both 4 : 8 and 5 : 10. An inverted ratio will reverse the terms, so it can be said, 8 : 4 is the inverse of 4 : 8, and the invariant ratio now is 2 : 1.

Few traders truly grasp the concepts of ratios and become bound by this limitation from doing more advanced work. As example, standing between the two-termed ratio and the four-termed proportion is the three-termed mean, in which the middle term is in the same ratio to the first as the last is to it. The geometric mean between two numbers is equal to the square root of their product. Therefore, it can be said, the geometric mean of 1 and 9 is $\sqrt{(1 \times 9)}$, which equals 3. This geometric mean relationship is written as 1 : 3 : 9 or inverted, 9 : 3 : 1. It can also be written more fully as a continuous geometric proportion where these two ratios repeat the same invariant ratio of 1 : 3. Thus, 1 : 3 :: 3 : 9. The 3 is the geometric mean held in common by both ratios. This is the interlacing mathematical glue that binds them together.

The Pythagoreans called this a three-termed continuous geometric proportion. The Golden Ratio has a three-termed continuous geometric proportion. As viewed in Figure 1.2, the Golden Ratio is the ratio that results between two numbers A and C when a line is divided, so that the whole line, AC, has the same ratio to the larger segment, AB, as the larger segment, AB, has to the smaller segment, BC. AC : AB :: AB : BC. The geometric mean held in common by both ratios is B or 0.618. As B is the geometric mean of A and C, it can be written as A : B : C. We traders will use this ratio to define the mathematical grid a market is forming to build future price swings and pivots.

Pythagoras

Pythagoras (570–490 BC) was born in Ionia on the island of Samos in Greece. He eventually settled in Crotone, a Dorian Greek colony in southern Italy in 529 BC. Pythagoras's thought was dominated by mathematics, and the Pythagorean philosophy can be summarized as: Only through mathematics can anything be proven, and by mathematical limit the unlimited can take form. The thinking of the Pythagoreans was exceedingly advanced for their time in many ways. As a simple example, they knew there must be a substance in the air to allow sound to travel from one person's mouth to another's ears. They knew nothing about atoms or kinetic energy, but already had the conceptual idea. Pythagoras's school gradually formed into a society or brotherhood called the Order of the Pythagoreans. Growing in wealth and power, Pythagoras[2] was murdered when their meetinghouse was torched. The remaining Pythagoreans were scattered across the Mediterranean. Pythagoras wrote nothing down and none of his followers' original papers on mathematics survived. However, after the attacks on the Pythagoreans at Crotone, they regrouped in Tarentum in southern Italy. Collections of Pythagorean writings on ethics[3] show a creative response to their troubles.

In a biography of Pythagoras written seven centuries after Pythagoras's time, Porphyry (AD 233–309) stated the Pythagoreans were divided into an inner circle called the *mathematikoi* ("mathematicians who study all") and an outer circle called the *akousmatikoi* ("listeners"). Pythagoras's wife Theano and their two daughters led the *mathematikoi*

FIGURE 1.3   Woman Teaching Geometry 1309. It is rare to find a woman teaching Geometry, one of the sacred sciences, as depicted in this painting from 1309. It shows her teaching a group of young monks.

after Pythagoras's death.[4] The early work of the Pythagoreans on the Golden Ratio might have been lost if it had not been for a woman: Theano, who was a mathematician in her own right. She is credited with having written treatises on mathematics, physics, medicine, and child psychology, although nothing of her writing survives. Her most important work is said to have been a treatise on the principle of the golden mean. In a time when women were usually considered property and relegated to the role of housekeeper or spouse, her work was in keeping with Pythagoras's views that allowed women to function on equal intellectual terms in his society.

Most of the concepts developed by the Pythagoreans gave credit to Pythagoras himself. So it is hard to unravel his specific accomplishments from his followers. But Aristotle, in his *Metaphysica*, sums up the Pythagorean's attitude toward numbers:

> The so-called Pythagoreans, who were the first to take up mathematics, not only advanced this subject, but saturated with it, they fancied that the principles of mathematics were the principles of all things.[5]

The Pythagoreans knew just the positive whole numbers, zero, and negative numbers, and the irrational numbers didn't exist in their system. Despite this limitation about irrational numbers, Pythagoras[6] is believed to have discovered the Golden Ratio through sound—the sound of two hammers of different weights hit an anvil producing a harmonic pitch of divine perfection. This fact, that the origin of the Golden Ratio was discovered through a harmonic ratio, will have implications in the final chapters of this book. It will change how traders should think and apply the Golden Ratio to analyze the markets.

While accounts of Pythagoras's travels differ, historians agree he traveled to many countries to study with the masters of his time. Many historians believed he was an initiate of the ancient Egyptian priests, as was Plato and Euclid. But why is this of interest in the business of trading today's markets? The ancient priests of Egypt believed that only a student well versed in the ancient quadrivium—the study of seven disciplines that included arithmetic, geometry, music, and astronomy—would be able to solve the problems they faced in their life. This philosophy for attaining wisdom is no different than what is needed to understand the full potential of the Golden Ratio applied to our global market puzzles today. It is only through a study of ratios can one understand the concepts of proportion, symmetry, harmonics, and rhythm. Many traders erroneously assume that markets move in fixed intervals within linear dimensions. *Markets do not move in such a manner.* Future market price swings will expand or contract in measurable ratios derived from *multiple prior price swings.* Markets are living breathing entities, and the industry is stuck with outdated assumptions and applications that we must reevaluate. In order to move forward beyond the elementary, we must revisit the teachings of the past so we can reprogram our thinking.

It is generally accepted that Plato offers the world the earliest written documentation of the Pythagoreans.[7] In 387 BC, Plato founded an academy in Athens, often described as the first university. Plato loved geometry also. His school included a comprehensive curriculum of biology, political science, philosophy, mathematics, and astronomy. Over the doors to his academy were the words αγεωμερητος μηδεισεισιτω, meaning "Let no one destitute of geometry enter my doors."

It was long thought both Pythagoras and Plato were by-products of Egyptian schooling. No proof to this regard was available until recently. The highly respected Egyptian Egyptologist Dr. Okasha El-Daly has released in his book, *Egyptology: The Missing Millennium,*[8] proof that the Greek philosopher Pythagoras had been a student of the carefully guarded wisdom of the ancient Egyptian high priests. Until now, this was only suspected to be true. During a dinner in Sharm El-Sheikh, Egypt, with Dr. El-Daly, he commented to me that the ancient Egyptians had such high regard for Pythagoras that El-Daly found Egyptian papyrus in the British Museum vaults documenting the construction of a tomb chamber within the Pyramid of Menkaure.[9] This is the third pyramid on the Giza plateau and the smallest of the three. Pythagoras did not die in Egypt and is not buried there, but the references to these preparations for an afterlife were thought unheard of except for the Pharaohs and immediate family members themselves. Why was Pythagoras so highly revered?

Here's the heart of what is of interest to us today. The Pythagoreans believed all things were connected. Movement formed vibrations that brought harmonic associations to all things. (Translated to *all* pivots within our price data.) It was thought that only through arithmetic could one prove a universal truth. You might say that the Pythagoreans were the first to consider the law of vibration, a concept of great interest to market analysts using the methods of W.D. Gann. Gann is the trader/analyst who correctly called the great crash of 1929 in his annual report written in 1928. Not only did he call the decline, he also recognized the exact bottom one month after the market low. You will begin to understand the law of vibration in the final chapter and why it is of interest to us.

It is useful to know that Gann was also trained in these ancient studies, and the concepts are still valued by Freemasons today as they seek enlightenment and truth. Gann attained the second-highest order of the Scottish Rite, and as a thirty-second degree Freemason he believed similarly to the Pythagoreans that *arithmetic is pure number, geometry is number in space, music is number in time, and astronomy is number in space and time.*

An invisible thread connecting all these fields of study is the ratio 1.618 (Phi) or its inverse reciprocal 0.618 (phi),[10] which proves to be more valuable in our applications of market analysis. These early concepts of the Pythagoreans unlock how multiple market pivots can accurately define the next price swing or inflection price levels that the market will respect. The underlying premise is to understand that markets grow and contract to universal constants that are predictable, measurable, and present in all market time horizons. Through Fibonacci analysis and the ratios of Phi (1.618) and its reciprocal phi (0.618), we can prove markets follow the laws of nature and we in turn must work with these underlying principles rather than force our will on the price data with erroneous assumptions.

Fibonacci

The Fibonacci number series: 0, 1, 1, 2, 3, 5, 8, 13, 21, 34, 55, 89, 144, 233, 377, . . . is both additive, as each number is the sum of the previous two, and multiplicative, as each number approximates the previous number multiplied by the golden section ratio. The ratio becomes more precise as the numbers increase. Inversely, any number divided by its smaller neighbor approximates 0.618 or phi $(1/\phi)$.

While traders with rudimentary knowledge know they can add, subtract, divide, and multiply Fibonacci ratios to obtain another Fibonacci ratio,[11] they often do not apply this knowledge in their chart analysis. As an example, traders during the Nasdaq crash in 2000 and 2001 would have had no means to determine a price target low if they only knew how to calculate price support levels by subdividing the extreme price high and historic low into the ratios of 61.8 percent, 50 percent, and 38.2 percent.

In **Figure 1.4**, we see the monthly chart for the Nasdaq Composite Index. Range A begins at the price high and ends at a price low significantly higher than the market low. The price low selected in 1998 was a significant decline, but was not the historic low, which would have made the problem we are about to discuss worse. Traders who simply subdivided the range as defined by range A would witness the market falling through the 61.8 percent support zone and not know what to do to identify new targets. A retracement of 61.8 percent is the decline that occurred relative to the full price range selected. Traders and analysts could further subdivide the range between the 61.8 percent level of A and the price low first selected to obtain additional Fibonacci ratios.[12] They in turn could find additional support levels by dividing the range marked C into the ratios 38.2 percent, 50.0 percent, and 61.8 percent. We could further subdivide range D. Each of these additional subdivisions shows that the Nasdaq Composite respected the ratios identified. Any Fibonacci ratio that is added, subtracted, multiplied, or divided will produce another Fibonacci ratio. Many traders know this, but few know how to apply it. Then I raise the question, should the key reversal spike high into the market high have been truncated when the start of the range was defined? This question will be answered and many more as these chapters unfold.

The key observation to draw from the Nasdaq chart in Figure 1.4 is that each price range identified—A, B, C, and range D—is subdivided into three ratios, 38.2 percent, 50 percent, and 61.8 percent, relative to its starting and ending price levels. The calculations that result from subdividing each range are entirely dependent on the range first selected, and these ranges require thought as they become critical to our success. We will not be starting at the price high and then using the extreme price low as favored by our industry to apply

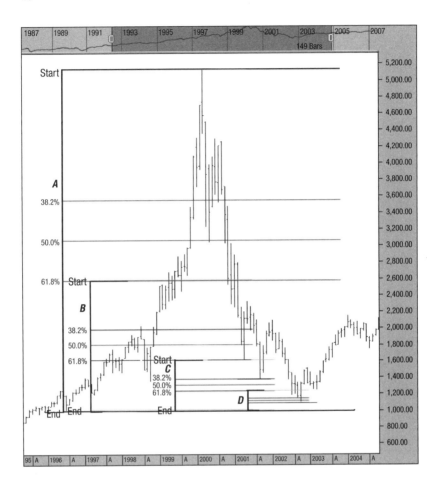

FIGURE 1.4   **Nasdaq Composite Index Monthly Chart**

Connie Brown, www.aeroinvest.com. Source: Copyright © 2008 Market Analyst Software

Fibonacci analysis. Figure 1.4 is a good introduction but more is involved. The correct ranges to use will be addressed in detail within the next chapter when we focus on the concept of market expansion and contraction.

Leonardo Fibonacci (circa 1170–1250) was not the first to document the Fibonacci number sequence. That honor falls to ancient Hindu mathematicians.[13]

FIGURE 1.5   Leonardo Fibonacci—(Leonardo da Pisa), by Giovanni
Paganucci, in the Camposanto di Pisa Cemetary, Italy

Source: Private photograph collection of Connie Brown

Fibonacci might have lost his identity with the number sequence entirely had it not been for Édouard Anatole Lucas[14] (1842–1891), a French mathematician who *rediscovered* the Fibonacci sequence in the late nineteenth century. It was Lucas who attributed the number series to Fibonacci's book *Liber Abaci* (*Book of the Abacus*, 1202), thereby establishing the name of the numbers as the Fibonacci Number Sequence. Lucas is known for defining a formula to find the nth term of the Fibonacci sequence.

Liber Abaci demonstrates routine computations that merchants performed when converting currencies. Fibonacci showed the advantages of using the Hindu-Arabic number system compared with the Roman numeral system. In hindsight, Fibonacci made things much harder than they needed to be by frequently expressing fractions as unit fractions. As example: *One quarter and one third of a tree lie below ground, a total of twenty-one palmi in length. What is the length of the tree?*[15] Fortunately, Fibonacci's book made a huge impact, and as a result, we do not have to trade markets using Roman numerals. But sadly Fibonacci's studies through Mesopotamia have been lost. The library and museum of Mesopotamian artifacts vanished in the twenty-first century, when the Iraq National Museum in Baghdad was looted and burned after the 2003 Gulf War invasion. In Chapter 4, you will see a valuable artifact within the British Museum of great interest, but an Egyptian Egyptologist was near tears when she told me during my 2007 visit to Cairo that all the manuscripts and papyrus they had in Iraq are now lost forever.

CHAPTER NOTES

Abbreviations are references to the bibliography contained at the back of this book. As example, Leh, 8, 17 refers to the alphabetized bibliography code, Leh, which further describes: Lehner, Mark. *The Complete Pyramids: Solving the Ancient Mysteries.* London: Thames and Hudson Ltd., 1997.

1. The first clear definition of what later became known as the Golden Ratio was given around 300 BC by the founder of geometry, Euclid of Alexandria. Euclid defined a proportion derived from a simple division of a line into what he called its "extreme and mean ratio." In Euclid's words:

> A straight line is said to have been cut in extreme and mean ratio when, as the whole line is to the greater segment, so is the greater to the lesser.
>
> —HEATH

2. Hall, 191–223; see also Taylor.

3. Taylor.

4. Hall; Taylor.

5. Aristotle, *Metaphysica*, 1–5.

6. Stroh, 82–85.

7. Robin.

8. El-Daly, 60.

9. Mad, 5. The story of phi begins in the fourth dynasty of Egypt, about 2500 BC. The last of the original great Seven Wonders of the Ancient World are the extraordinary pyramids of Egypt. The Great Pyramid of Khufu (called Cheops by the Greeks), the Pyramid of Khafre (Chephren), Khufu's grandson, and the smallest of the three called the Pyramid of Menkaure, are located on the Giza plateau near Cairo.

Everyone agrees on the meticulous precision with which the pyramids were built. Khufu is oriented to within three feet six inches of an arc from true North. In addition, it is well known that Khufu contains the phi proportion between its sides and base. With only rudimentary arithmetic and only ropes or sticks for measuring, this seems like an improbable result. But we will study this geometry further in Chapter 4.

Khufu has been measured repeatedly since the seventeenth century AD, each time with greater accuracy. A right angle could be constructed by folding a rope with thirteen equally spaced knots, such that three spaces between knots form one side of a triangle, four spaces form another side, and five spaces form the hypotenuse. Several researchers have found murals and papyruses that show knotted ropes. Several pyramids were built on the proportions of the three-four-five triangle. The height, half-base, and slant angle of Khafre have been measured at 143.5 meters, 107.5 meters, and fifty-three degrees ten feet. The ratio of these, $143.5/107.5 = 1.33488$, is very close to $4/3$ as verified by the angle. (Leh, 7) This ratio of 1.33 will take on greater significance later in the book.

10. *Merriam-Webster's Collegiate Dictionary: Tenth Edition* shows that Phi (ϕ) is pronounced, "fi" like *fire*, and not *pie*. The reciprocal phi or $1/(\phi)$ is pronounced, "fee" like *bee*.

11. Fibonacci ratios can be added, subtracted, multiplied, or divided and the result will always be another Fibonacci ratio. As examples:

$$0.618 \times 0.618 = 0.382 \qquad 0.618 - 0.382 = 0.236$$

$$0.382 \times 0.618 = 0.236 \qquad 0.382 - 0.236 = 0.146$$

$$0.236 \times 0.618 = 0.146 \text{ etc.} \qquad 0.236 - 0.146 = 0.090$$

12. Ibid.

13. Gies, 57, 110, and Dunlap, 35. The Italian mathematician born 1170 AD in Pisa, Italy, was commonly known as Fibonacci, which was a shortened form of Filius Bonaccio (son of Bonaccio). Fibonacci's book *Liber Abaci* (*Book of the Abacus*) introduced in 1202 the Hindu-Arabic system of numbers to Europe. Fibonacci states in his introduction that he accompanied his father Guilielmo, on an extended commercial mission in Algeria with a group of Pisan merchants. There, he says, his father had him instructed in the Hindu-Arabic numerals and computations. He continued his studies in Egypt. (Cooke, 289) (Fibonacci was recognized as a great mathematician, but his solution to the rabbit problem was overlooked when he determined the number of immortal pairs of adult and baby rabbits each month over a one year interval revealing the Fibonacci number sequence 0, 1, 1, 2, 3, 5, 8, 13, . . . and so on.)

See also Parm and Parm, 85, and Pin, 45. In the classical period of Indian mathematics (400 AD–1200 AD), important scholars like Aryabhata, Brahmagupta, and Bhaskara II made early contributions to the study of the decimal number system, zero, negative numbers, arithmetic, and algebra. Trigonometry was introduced into ancient India through the translations of Greek works, further showing that Fibonacci applied the number sequence to solve the rabbit problem, but he did not discover it.

14. Hark, 276–288.

15. Cooke, 289.

CHAPTER 2

The Concept of Market Expansion and Contraction

FIGURE 2.1 Iceland Low

Source: Jacques Descloitres, MDOS Rapid Response Team, NASA/GSFC

To understand market expansion and contraction, you need to understand the differences between ratios, means, and proportion. In this chapter, the concepts of market expansion and contraction will be evaluated within bar charts, and then you will learn how to develop the unique support and resistance price grid a specific market uses to create its next market swing. The voids between the levels of support and resistance help you work smarter with oscillators and with other technical tools as well. So the best place to start is with a quick review of the terms *ratio*, *mean*, and *proportion*.

Ratio: A ratio between two numbers *a* and *b* is a : b or a/b.

The inverse ratio between the two numbers *a* and *b* is b : a or b/a.

The Fibonacci number series is created by adding two numbers together to form the next in the series, 0, 1, 1, 2, 3, 5, 8, 13, 21, 34, 55, 89, 144, . . . and so on. When you want to know the relative relationship between two different numbers, you divide one by the other. If you divide 144 by 89, the result is a ratio of 1.6179. You can round this result to 1.618. If you divide 89 by 55, you obtain the ratio 1.61818181818. . . . In fact, if you divide the higher numbers after 233 by their lower number in the number sequence, you find a near exact 1.61805233 result, as an example, 233 divided by 144. Each pair of numbers will have the same corresponding relationship between them. So you know 1.618 is important but not why just yet.

If you divide 55 by 89, you produce a reciprocal relationship between the two values of 0.61797, or rounded 0.618. If you divide 89 by 144, you find a similar result of 0.61805. So 1.618 (Phi or φ) and the reciprocal 0.618 (phi, 1/φ) are clearly ratios of significance when you compare two nearby numbers in the Fibonacci sequence.

If you consider alternate relationships between two points, such as 21 divided by 55, you obtain the ratio 0.3818181818 or 0.382. This will be another ratio of great interest to you when you start analyzing markets.

Mean: If you want to find the mean *b* between *a* and *c*, you may consider three common approaches. There are others but you will not be encountering them.

and c is b = (a + c)/2.

and c is b = $\sqrt{ac}$.

nd c will equal the following:

(2ac)/(a + c).

the arithmetic mean, and it is usually

t say, "mean." The geometric mean is

owth. The growth of sales for a business

in a rally uses the geometric mean.

eometric mean as opposed to arithmetic

geometric mean in general?

elevant any time several quantities add

The arithmetic mean answers the ques-

tion, if all the quantities had the same value, what would that value
need to be in order to achieve the same total?

In the same way, the geometric mean is relevant any time several
quantities multiply together to produce a product. The geometric mean
answers the question, if all the quantities had the same value, what
would that value need to be in order to achieve the same product?

For example, suppose you have an investment that earns 10 percent
the first year, 60 percent the second year, and 20 percent the third
year. What is its average rate of return? It is *not* the arithmetic mean,
because what these numbers indicate is that on the first year your
investment was *multiplied* (not added to) by 1.10, on the second year it
was multiplied by 1.60, and the third year it was multiplied by 1.20.
The relevant quantity is the geometric mean of these three numbers.

The question about finding the average rate of return can be
rephrased as, by what constant factor would your investment need
to be multiplied by each year in order to achieve the same effect as
multiplying by 1.10 one year, 1.60 the next, and 1.20 the third? The
answer is the *geometric mean* (1.10 × 1.60 × 1.20). If you calculate this
geometric mean, you get approximately 1.283, so the average rate of
return is about 28 percent (not 30 percent, which is what the arithmetic mean of 10 percent, 60 percent, and 20 percent would give you).

Any time you have a number of factors contributing to a product
and you want to find the "average" factor, the answer is the geometric
mean. The example of interest rates is probably the application most
used in everyday life.

It is known that the geometric mean is always less than or equal to the arith-metic mean (equality holding only when $A = B$). The proof of this is quite short and follows from the fact that ($\sqrt{A} - (\sqrt{B})^2$) is always a non-negative number. In the latter pages of this book, it will be important to know the difference, as you will begin to look at harmonic propor-tion within the markets.

The harmonic mean is used to calculate average rates such as distance per time, or speed. An analyst might consider the use of a harmonic mean when a third axis incorporates harmonic frequen-cies derived from ratios the market swings produce. This will be a new area of geometry for our industry discussed near the end of this book, but it is essential the concept of proportion is accurate in your mind.

Proportion: If we consider Plato's[1] request to make an uneven cut, we might realize an even cut would result in a whole: to a segment ratio of 2 : 1. The two equal segments would be 1 : 1. These ratios are not equal 1 : 1 :: 2 : 1, and so they cannot present a proportional rela-tionship. There is only one way to form a proportion from a simple ratio, and that is through the golden section. Plato wants us to dis-cover that only one special ratio exists such that *the whole to the longer equals the longer to the shorter*. (See **Figure 2.2**.) He knew this would result in the continuous geometric proportion found in nature with-out violating his secret oath.

Why did Plato use the division of a line rather than give us num-bers to work from? The reason is the Fibonacci numbers give an irra-tional number that cannot be expressed as a simple fraction. Only by solving this problem geometrically can we discover that the longer segment will always have a value of 1.6180339, Phi (ϕ) relative to the whole. How this applies to chart analysis is this will be true regardless of what pivot points we use to define the range. The lesser segment of the line will always be 0.6180339 or phi (1/ϕ). The mean value is 1. The Pythagoreans attached great meaning to this mean value of 1,[2] as it represents the unity that binds all living forms. When you see books on the Golden Ratio subdividing fish, horses, flowers, and shells, the math ratios and proportions are surprisingly similar. But what is criti-cal in market analysis is to know what internal structures are sig-nificant to create the price range to be subdivided. There are clues

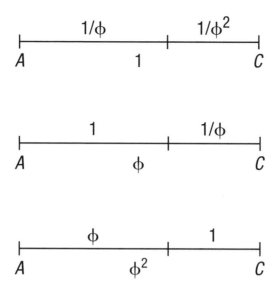

FIGURE 2.2   The Golden Ratio Proportions

everywhere to help us select the right ranges to subdivide, but they are rarely the most obvious price highs and lows. It is only through divine proportion can we see the unity of all in nature. Soon you will see markets abide by the same unity.

There are two kinds of proportions to consider. The first is *discontinuous* (four-termed) such as 4 : 8 :: 5 : 10 or a : b :: c : d. This example has an invariant ratio of 1 : 2. The second kind of proportion is *continuous* (three-termed) or a : b :: b : c, which equals a : b : c, where b is the geometric mean of a and c. In market charts, you always select two market pivot price points, one being a price low and the other being a high or vice versa to subdivide the price range. The 61.8 percent ratio will be the farthest ratio from the starting point of the range. In the weekly chart for Caterpillar Inc. (CAT) stock in **Figure 2.3**, the distance from the price swing high for CAT to the 61.8 percent retracement level and the distance from the price low at *B* to the 38.2 percent retracement have equal lengths. It can be said of Figure 2.3 that the price high is to 61.8 percent as the price low is to 38.2 percent.

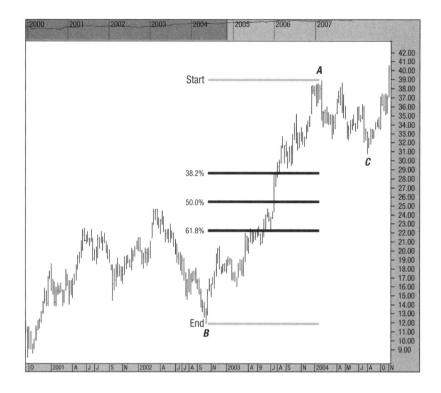

FIGURE 2.3   **Caterpillar—Weekly Chart**

Connie Brown, www.aeroinvest.com. Source: Copyright © 2008 Market Analyst Software

Plato went one step further by referencing an extended continuous geometric proportion such as 1 : 3 :: 3 : 9 :: 9 : 27. This proportion has an invariant ratio of 1 : 3 or 1/3. This ratio will have specific implications for how you can project future prices when the market is breaking into new highs. It also has an influence on where you should make that projection *and often it is not from a price low that ends an old trend*. As these concepts are used in the analysis ahead, you will find much of what you will do is visual. It will be applied geometry, and the actual math behind the ratios and proportions will be less difficult than you may think now.

The chart in Figure 2.3 shows how most traders create Fibonacci ratios within a bar chart. Actually, this chart begins to deviate from what most traders do slightly. Most traders will define the range by starting at the extreme price low of a major swing and then select the price high. You will soon see why you must never do this. After teaching numerous seminars this next point seems to catch everyone in application. May I suggest you underline or, even better, write this down on a separate note pad and attach it to your computer:

> When you want to find a price retracement level for support (under the current market price), *start with a price high and move your cursor down* toward a price low to define the range to be subdivided. (Also, the 38.2 percent, 50 percent, and 61.8 percent subdivision results *must* fall below the current price.)

Most quote vendors have taught you to start at the bottom because they draw their lines on your screen incorrectly by extending the lines away from the y-axis and the new price data. If they are wrong, by drawing the subdivided results behind your cursor rather than forward towards the most recent price bar,[3] the problem can be easily fixed. Change your default settings so that the line ratios are drawn across the entire computer screen. If your vendor's default setting includes more ratios than 38.2 percent, 50.0 percent, and 61.8 percent, remove the extra ratios. As an example, you will not use 75 percent.

When you want to find the resistance price level (over the current market price), you must *start with a price low and move your cursor toward a price high.* (The resulting subdivisions of the range *must* be above the current market price.)

Why? If you select your price range in the opposite manner, the method will force you to remain a beginner, as you will never be able to consider the internal proportions forming within the range of any price swing. This will become clearer as this chapter develops.

In Figure 2.3, a price high is defined at pivot *A* for Caterpillar, and a price low at pivot *B*, as most traders have been taught. However, Caterpillar never realized the first support level within this subdivided range near $29 (the 38.2 percent retracement) when the pullback occurred to the price swing low at pivot *C*. What happened? Fibonacci doesn't work, some will say. They would also be right if all they ever do is take a price high and price low to define the range in this manner and think they have a viable target price. The reason this calculation is off is this market is expanding.

Caterpillar was preparing to explode upwards in price as you see in **Figure 2.4**. This is the same weekly chart for Caterpillar, but the retracement ratios to define a price target for the trade entry level have been taken from a different swing range. In this chart, the high at point *A* is the same price selected in Figure 2.3. But point *B*, marking the low of this new range, is *one swing higher* than selected for the calculations in chart Figure 2.3. W.D. Gann stated in his stock course[4] that he often found the secondary swing away from the actual bottom, or the secondary high after the end of a trend, to be of greater value than the

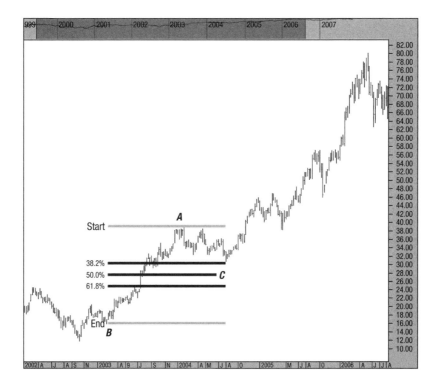

FIGURE 2.4   Caterpillar—Weekly Chart

Connie Brown, www.aeroinvest.com. Source: Copyright © 2008 Market Analyst Software

actual price that ends the prior trend. By moving the low of the range first selected up one swing, we find the market retracement to point C trades at our calculation for a 38.2 percent support target. I have found this to be true and will show you how markets give us internal price clues that tell us when we should make adjustments like this and when we should not. But first we need to build a stronger foundation on the concept of market expansion and contraction, because this is the reason why the calculation in Figure 2.3 did not let a trader enter the market at his target.

Every reader has likely seen the stunning curvature that forms in the ivory shell from the Indian Ocean called *Nautilus pompilius* (**Figure 2.5**). This sensuous curve can be found in many other forms of nature such as hurricane clouds (see Figure 1.1), DNA nucleotides, sunflowers, the solar system, and in chemistry the uranium oxide compounds U_2O_5, U_5O_8, U_5O_{13}, U_8O_{21}, and $U_{13}O_{34}$.

Opposing spirals of seeds in a sunflower generally appear as adjacent Fibonacci numbers, typically 55 : 34 (1.6176) or 89 : 55 (1.6). Scales of pinecones are typically 5 : 3 (1.666) or 8 : 5 (1 : 6). Artichokes display eight spirals going one way and five spirals the other. Pineapples have three spirals, often 8, 13, and 21 where 21 : 13 : 8 approximates ϕ : 1 : 1/ϕ (0.618 : 1 : 1.618). Nature pulses with cycles and rhythms that increase and decrease. Plato noted, "The way up and the way down are one and the same."[5] On the way up, nature uses additive and multiplicative relationships. On the way down, nature uses subtraction and division to diminish and create the logarithmic growth and decay cycles you see exemplified by the nautilus shell. The simple logarithmic growth spiral, or golden spiral, is derived from Fibonacci numbers.

When the spiral is logarithmic, the curve appears the same at every scale, and any line drawn from the center meets any part of the spiral at exactly the same angle for that spiral. But the bigger question is this, how do you make money from a curve that wraps round and round when your market chart has the current data on the right side of your computer screen and the oldest data is dropping off the left side of the screen?

Every book I have seen has these stunning images and authors seem to think this relationship between the curve and a two-dimensional chart on paper is intuitively clear. Well, it is not. In fact, it took me some

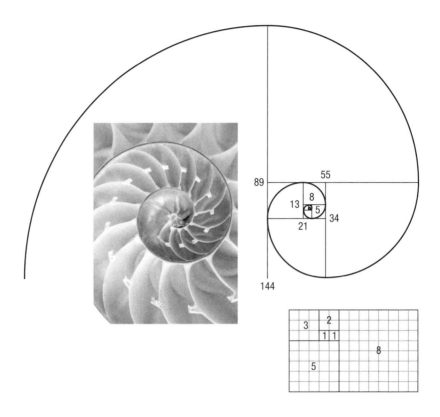

FIGURE 2.5   **Nautilus Pompilius Shell Compared to the Golden Spiral**

Source: Private photograph collection of Connie Brown

time to really get it. It's not the Fibonacci numbers that everyone is talking about that will turn you into a market guru; *it's all about the ratios that form between these numbers that will take you to the bank.* Depending on how these ratios fall within our market data, the ratios that cluster together can even warn us how fast the market will move towards the target. You will see examples in Chapter 6 when we consider market character between target zones.

Look very, very closely at the mathematical representation laid over top the nautilus shell in **Figure 2.6**. When you study this shell as an overlay with *the theoretical Fibonacci squares* that create the spiral, you can see more easily that the shell curvature departs from the ideal

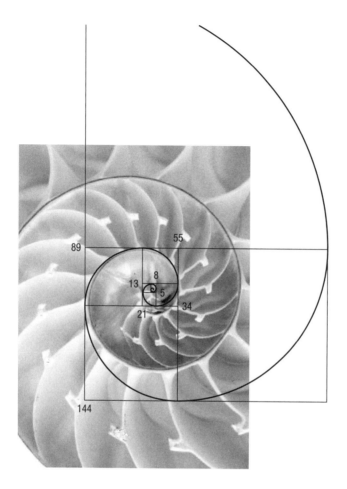

FIGURE 2.6   **The Nautilus with the Golden Spiral**

Source: Photograph and artwork Connie Brown

mathematical model for perfection. Side by side in the first diagram (Figure 2.5), they looked the same but they are not.

The curvature of the shell tracks the math model to perfection closely up to the line for 21 units in length. Then the curvature of the shell departs the math model by showing a tighter growth spiral than the ideal math projection. *This contraction is reality. The math model is theory.* Anyone who has ever calculated a Fibonacci ratio by slapping a mouse pointer on a price high and defined the extreme end of the range at the price low, has missed the boat about what these ratios mean within the context of an expanding or contracting growth cycle. You may have thought Fibonacci doesn't work in the markets as a result. But Fibonacci ratios work no differently in markets from how they unfold in nature. There is nearly always a contraction or expansion factor at play within the market price swings, and this forces you to depart from making a theoretical projection. If you learn to work with the laws of nature and change your expectations that markets will mirror theoretical perfection, you will find that all markets in any time horizon will respect your targets more often.

Market Symmetry, Expansion, and Contraction

Staying with the weekly chart for Caterpillar in **Figure 2.7** allows us to measure an expansion cycle within this stock. In Figure 2.7 are three market ranges (*d*, *e*, and *f*) marked with vertical lines on the right side of the chart. The market swings for Caterpillar have been marked with light gray within the chart. Focus first on the shortest range, *d*, found in the middle of the chart. A double vertical line is used for range *d* to help your eye focus on the first range drawn.

The purpose is to see if this stock is developing swings that are expanding, contracting, or remaining symmetrical. Range *d* is the strongest part, or most forceful swing, of the move up in this chart. I am selecting a price high and low that allows the 50 percent line to fall within the strongest single bar within this swing. (If you know the Elliott Wave Principle the bar to study is Wave 3 of iii.) Line *A* is the result of bisecting range *d* equally, giving the 50 percent value of the range. Now move up to range *e*, where one corrective swing and one rally swing have been added at either end of range *d* to expand the first range used.

Range *e* defines the 50 percent division at line *B*. This line is slightly higher than line *A*. This means the market has expanded its proportional swings in *e* relative to range *d*. Now consider price range *f* created from a range enclosing the price high and price low in this chart. The 50 percent line at *C* has shifted slightly higher than line *B* in the second range. The expansion cycle in this market is increasing. If a trader only used

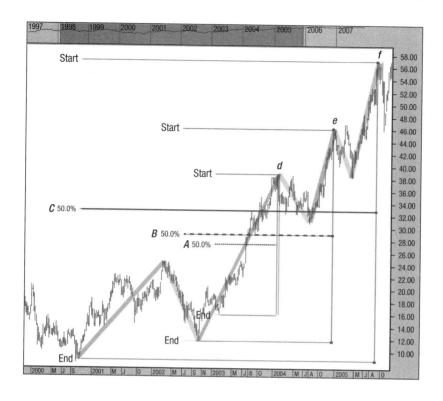

FIGURE 2.7   Caterpillar (CAT)—Weekly

Connie Brown, www.aeroinvest.com. Source: Copyright © 2008 Market Analyst Software

Fibonacci retracements by taking a price high and low, the trader would have to chase or never get in the market because the market reverses before reaching their targets.

When a market is contracting, the 50 percent lines within multiple swing ranges will fall below the 50 percent lines. It means the final swings are likely contracting as the market loses momentum.

The only time a trader will have success by using a Fibonacci ratio derived by taking the extreme market high and low is when all the 50 percent lines fall on top of one another. In this case, the market is developing swings in a symmetrical proportion, and the upper swings likely mirror the lower swings. Markets spend more time developing expansion and contraction proportional swings than they do symmetrical. So a trader with basic understanding of how to create Fibonacci retracements will only have marginal success. The other problem is if you start to define your ranges from the wrong direction. You will always be forced to assume the market is forming symmetrical swings because you cannot see any other result. All ranges to define support will start from the same high. All ranges to create resistance will start from the same price low. But do not assume the actual market extreme is the correct starting point.

It is hard to abandon old ideas or ways of applying a method, so another example is offered. In this example, the market will show us it is contracting. In **Figure 2.8**, we see a weekly chart for Centex

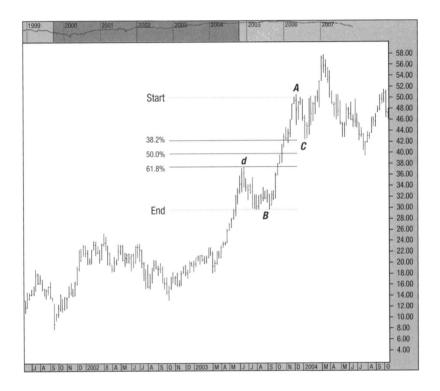

FIGURE 2.8   **Centex Weekly Chart**

Corporation (CTX). Consider the chart with no data created after the high marked point *A*. We want to determine a retracement price target to enter this stock. From point *A*, we start from a high and work down to a low to define our range. The end of this strong swing up is at price level *B*.

Level *B* has something extra you should notice: it marks a double bottom that is a directional signal. After the range is subdivided, look for internal milestones within the price data to confirm you are on track. In this case you are, because the price high at *d* is exactly under the 61.8 percent line. Therefore, you can see the market has respected this ratio before. It is important to say that you should not pick level *B* because *d* fits. Only look back after the range is set. Confidence you selected the correct range occurs as you study numerous internal milestones within the price data, which I will discuss shortly.

The simple retracement calculation at the 38.2 percent line holds the market perfectly at price low *C*. This corrective pullback confirms you made the right calculation, but it comes after the fact. Another problem develops with just one range selected; you cannot tell where the market is going next or where the market should not go. You have not defined where to put your stops either. So there is much more ahead for us to discuss.

Centex Corporation (CTX) is displayed again in **Figure 2.9**. A new range has been selected when a new high develops at point *A*. The end of the range is marked *B* and the 50 percent line falls within the middle of the strongest swing up. However, this time there is a big difference from the example for Caterpillar. This time the 50 percent line subdivides the larger range *below* the 50 percent line that marks the middle of the smaller internal range discussed in Figure 2.8. Centex is showing that the weekly data is beginning to contract. If point *B* truncated the small key reversal spike by bringing the bottom up to the low of the bar behind the spike, we would have found the key reversal at point *C* was off by the same amount. So I used the price low of this swing low at *C* to confirm this was the correct range to select.

I will show you more about how to select the ranges and what internal milestones or clues to look for, so you can have confidence that you are doing it correctly. The key and concept people miss when analyzing the nautilus shell is that the logarithmic spiral is the

FIGURE 2.9   Centex Weekly Chart

Connie Brown, www.aeroinvest.com. Source: Copyright © 2008 Market Analyst Software

same at every scale, *and any line drawn from the center meets any part of the spiral at exactly the same angle for that spiral.* You will see that markets give you the points you should use to create your analysis. The same markers will form in every market, regardless of time horizon, but you have to know what points to use to subdivide a range and create your ratios. Markets often ignore the price bar when trader sentiment runs amok forming a key reversal. The markets give little weight to the fact someone was foolish enough to enter a market order during the lunchtime doldrums of the day allowing a market maker to run the stops. I will give you a consistent blueprint for uncovering individual price grids hiding within price data. These points have proven to be reliable over twenty years of trading in trending or volatile market conditions.

Figure 2.10 displays a 3-day chart of Centex Corporation. This chart clarifies why Figure 2.9 showed a market beginning to contract

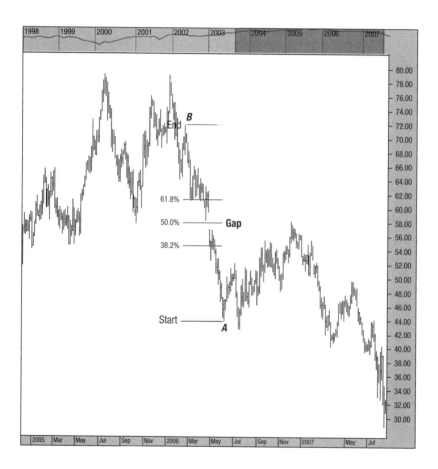

FIGURE 2.10   **Centex 3-Day Chart**

Connie Brown, www.aeroinvest.com. Source: Copyright © 2008 Market Analyst Software

within its internal proportions during the rally and for good reason. A major top in the home building sector was approaching. So we understand markets expand and contract and it raises the question, how do we select a range to be subdivided if the market proportions are shifting in expansion and contraction cycles all the time? The market will show us how it is using specific bars to bind the proportions, allowing us to calculate meaningful support or resistance targets using Fibonacci ratios.

Figure 2.10 contains the first internal milestone that always helps define the start and end of a range. Always align a gap near the 50 percent retracement line. In this chart the market has declined sharply to price low *A*. From *A*, the stock has a bounce and then marginally breaks below *A*. What is the correct low to use? We want to find a resistance target over the current market price *so we must start at the price low.* If we start at price low *A* and drag our mouse up to the price high at *B*, we will see the 50 percent line falls in the middle of the gap. If you started from the actual market high and dragged your mouse down to the low, you would never see this internal milestone form. That's why you must start from a price low and consider key swing highs within the range of the decline to find resistance. If the price low selected was the key reversal just past point *A*, you would have discovered the range high at point *B* would draw the 50 percent line right on the price highs forming on the bottom of the gap. In intraday trading, I often use this range for an early entry, but I know my stop has to be slightly higher. We will look at stop placement later in the chapter. So Figure 2.10 has introduced the first internal milestone the market gives you to begin to develop the price grid that the market will respect in future price swings. This is only the start.

Now we want to examine the data to see if the market has shown respect to the subdivisions within our price range in the past. By extending the subdivisions of the range in **Figure 2.11** to the left, we can see a few clues, but they are subtle and difficult for a first example. The corrective rally that follows into the 50 percent retracement seems to be confirmation, but keep in mind this only confirms that we took the correct range after the fact. We need confidence at the time we first define the range to subdivide.

FIGURE 2.11   **Centex 3-Day Chart**

Connie Brown, www.aeroinvest.com. Source: Copyright © 2008 Market Analyst Software

By extending the lines to the left in **Figure 2.12** and changing the look-back period of this chart to include data from 2003, we can see several points to discuss where the market has respected these levels in the past. *Keep in mind that the range was selected when we could not see this data.* The first set of clues that range *AB* has been correctly identified is the result of the market respect shown at pivots *e*, *d*, and *c*. If you had used the price low just after point *B*, all these key pivots would have been off an equal amount that the swing low falls under *B*. As we study

FIGURE 2.12   Centex 3-Day Chart

Connie Brown, www.aeroinvest.com. Source: Copyright © 2008 Market Analyst Software

the chart further, we see points *f* and *g* where strong moves began or were cut off. Point *f* will have great significance after the next example, but as an introduction to where the market truly starts a strong melt-down or thrust up, these internal milestones will always be important within the data. The point I felt would be hard to start with, when the chart in Figure 2.8 was viewed, is seen at point *h*. A back-and-fill pattern after *h* told me the selected range was correct, as the 61.8 percent retracement held the market over a period of many days. I will also notice when all the closes stay above (or below) such an internal line. In my experience, most traders in my seminar look only at the big picture and do not know how to look at the internals of the price data for valuable information. So I know this takes time and some work, because it is very different from how you likely read charts now.

I have been reading charts a long time using these methods and am very fast at it. The speed comes from knowing there are a few internal points that the market *always* uses as critical markers. Gaps are one type of milestone, but there are many others such as strong bars. I'll cover these next.

The decline of Centex continues to unfold, and you see the results in **Figure 2.13**. We sold the directional signal just behind point B and ran the trade into the lows near point A in this chart. How the targets for the decline were created will follow shortly, for now just study the entry levels at B and C into the corrective rally highs. Assume we unwound the trade near the lows at point A, and the indicators show that this market will still develop more of a decline.

Where do you get back in? Using the price low at point A to start your range, drag your mouse up to point B immediately. This is one of the major internal milestones or clues I was referencing. After the key reversal high just behind B in Figure 2.13, you see three bars showing the market could not exceed resistance at this level. (The mere fact that the bar highs all top near $49 is enough evidence to correctly call this area of resistance from the price data alone.) But from this small area of resistance, you see a strong break that moves the market towards $44 in only two bars. *Every time* you find the market starting strong trends like this in any time period, use the data to your advantage and define the end of the price range at this level and do not go on to select the price high.

Range AB is picked, and now we must ask if the three subdivisions within the range AB are major or minor resistance levels. To answer this, we must define a second range of different length, but the start of the range *must* be from the same low at A. Within the data, I see immediately the place where the next calculation must be made, as another strong bar shows a breakdown at b. The second range uses the same low as A so a marks the same start. The end of the range that we want to subdivide is b. We may not use *any* bar

FIGURE 2.13 Centex 3-Day Chart

Connie Brown, www.aeroinvest.com. Source: Copyright © 2008 Market Analyst Software

high that has been retraced in a corrective swing up. Therefore, all the bars that form behind $50 from the $40 low were retraced and cannot be used to mark the end of a price range. The first strong bar that developed is at *b*. Keep in mind that this calculation is being done before the swing forms from *A* to *c*. The entry target is where the two ranges form a confluence zone. Confluence forms on this chart where the 50 percent retracement aligns with the 38.2 percent retracement. Confluence zones form when different Fibonacci ratios come close or overlap from multiple ranges. Confluence zones are price inflection points of major support or resistance. Now we are really making progress because we can answer where the market is going to go next, when the market first starts to roll up from the lows at point *A*.

As a trader, you now want to know where the market should not go, so that you know where to put your stops if the rally continues higher than your confluence target zone at $44. The rally that develops and stops dead on the line at *c* is no accident. *Only look at your indicators as the market comes into the target confluence zone.* You will learn the market respects these confluence zones that form from ranges ending at strong internal bars, in every time horizon. Later as this method is developed further, we'll look at international global indexes to see how global strategies can be built with this analysis. But we need to stay with the basics for now.

We know where the market is going once *A* has been established. The market will target the confluence zone at *c*. Now we want to determine where the market should not go. To answer this question, we need a third range and the question to ask, is the pivot at *Q1* the right one to use or should you stick with the theme here and go straight up to *Q2*, to the strongest bar where the meltdown begins with conviction?

In **Figure 2.14**, I have temporarily removed the last range I added in Figure 2.13 so you can easily see the third range selected and why. Again you start the range from the same pivot low and end the range at the price high that falls between $55 and $56. Study the internal ratios that fall on the 61.8 percent, 50 percent, and 38.2 percent. You will find not a single price bar respects these levels. If the market does not respect the range you picked, it is not the grid the market is using to build its future swings so don't stay with it.

In **Figure 2.15**, the range has been adjusted so the new high aligns with the bar that starts this meltdown. I do not mean the bar that starts the trend, but the strong bar that is the world's point of recognition that the market is falling like a rock. In this price bar, we see prices fall from just over $56 to less than $53. Price points *m*, *n*, *o*, *p*, and *q* are the bars I pay attention to. My eye always travels from right to left looking at the more recent data and scanning back.

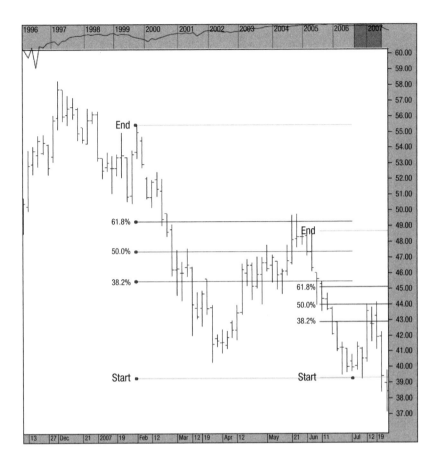

FIGURE 2.14 **Centex 3-Day Chart**

Connie Brown, www.aeroinvest.com. Source: Copyright © 2008 Market Analyst Software

These points all confirm the market is working off these proportions in some manner. In Figure 2.15, a range that was developed in Figure 2.14 was removed so you could see the details. Now the missing range needs to be redrawn so you have the full picture.

Figure 2.16 shows all three ranges and their subdivisions in the 3-day Centex chart. The confluence zone marked *x1* was the target to sell that was identified from the first two ranges. When a third range was added, a new confluence price zone forms at *x2*. We are armed

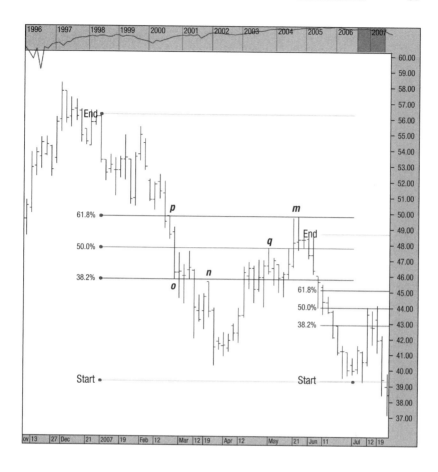

FIGURE 2.15   Centex 3-Day Chart

Connie Brown, www.aeroinvest.com. Source: Copyright © 2008 Market Analyst Software

with a lot of information right now. We know the target to reenter the market for another swing down is at confluence zone *x1*. We know the single line that formed just over this zone near $45 that is a 61.8 percent ratio from the first range selected is of less interest to us or to the market. We know the next confluence zone at *x2*, where a 50 percent and 38.2 percent ratio nearly overlap, is the price zone the market should not exceed. As a result, we know where to put our stop, which is just over the confluence zone *x2*. Since we have an

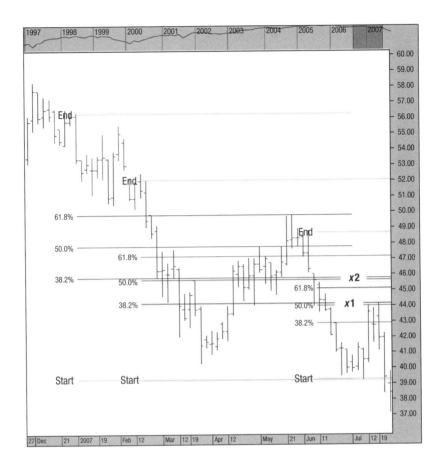

FIGURE 2.16 Centex 3-Day Chart—Two Fibonacci Confluence Zones

Connie Brown, www.aeroinvest.com. Source: Copyright © 2008 Market Analyst Software

entry and we have a precise exit level where we know we are wrong, we know what our risk-to-reward ratio will be. If we believe all trades should have at minimum a 1 : 3 risk-to-reward ratio, we would see we are risking a tad more than $2. Better to enter near $44 and exit just past $46. (When you enter stops, never use an even number in any market.) To keep a 1 : 3 risk-to-reward ratio, the market must move $6 away from $44. In Figure 2.16, this occurred, but we made the projection before the trade was established. To those who believe Fibonacci ratios are of little value; let them continue to do so, as those of us who work to develop these concepts will see bankable results.

In this major decline of Centex, we did not discuss how to develop targets for the meltdown. We covered how to create entry levels and stop placement, but methods to determine price targets for future swings must follow in the next discussion. I have not forgotten to answer the question, how do you make money from a spiral that turns round and round when you have to use two-dimensional charts? The geometry will be discussed in Chapter 4, and then developed further in the final chapters. The nautilus shell is in fact the most sophisticated geometric model you will encounter, as it has implications to traders in both the x and y-axes impacting both *price and time analysis*.

CHAPTER NOTES

1. Cooper, 1130–1132. See Plato *Republic* 509d–513e.

2. Cooper, 1224. Plato's Second Principle, known as the Indefinite Dyad, is sometimes called the Greater and the Lesser, and its relation to the golden section, ϕ. Socrates and Timaeus banter the importance of these relationships in Plato's work *Timaeus* using complex triangles. *Timaeus* shows how geometric form was used to argue logic. All of the Pythagorean symbols have important meanings tied to the arrangement and proportions within the geometric forms known as sacred geometry.

$$\text{Greater} = \phi$$
$$\text{Lesser} \;\; = 1/\phi.$$
$$\phi = \frac{\sqrt{5} + 1}{2} \approx 1.618.$$

Given the facts above, Plato revealed the Pythagorean belief that these relationships connected all:

$$\text{Greater} \times \text{Lesser} = 1$$
$$\text{Greater} - \text{Lesser} = 1$$
$$\text{Greater} \div \text{Unity} = \phi$$
$$\text{Unity} \div \text{Lesser} = \phi$$
$$\text{Greater} \div \text{Lesser} = \phi^2$$

3. CQG traders, in particular, will have difficulties. Change your default as described on page 21.

4. The W.D. Gann Stock Trading Course Collection.

5. Cooper.

CHAPTER 3

Support, Resistance, and Price Projections

IN THE LAST DISCUSSION, Figure 2.16 defined two Fibonacci confluence zones that formed within the 3-day Centex Corporation chart. Our purpose for defining these target confluence zones was to enter a short position in a developing trend. The intraday charts showed the market should not exceed the zone above, and we knew therefore to place stops over the next target zone. *Stops are not within the zone, but just above the zone.* Longer-horizon traders would have made the same calculations but might have needed to consider their stop exit level above one zone higher again. In Figure 2.16, we did not create the third confluence zone, but we will do so as these examples unfold.

As we know where to sell and where to put our stop, we need to know now where the market is going to complete a risk-to-reward ratio for our risk management needs and position size determination.

The two confluence zones derived from *different Fibonacci ratios* created in Figure 2.16 are seen passing through point *a* in **Figure 3.1** and extend across the chart. On the far left, the ranges used to create these zones are still visible. They have dots on the calculations. They can be deleted once a horizontal line has been added to mark the two confluence zones at 45.56 and 43.97, as they are not needed any longer. Both charts show the 3-day Centex Corporation data but Figure 3.1 now shows the market decline that formed after the short position entry at point *a*.

The question to answer was, how do we determine the price target if we sell into point *a*? The easiest method is to draw a box marking

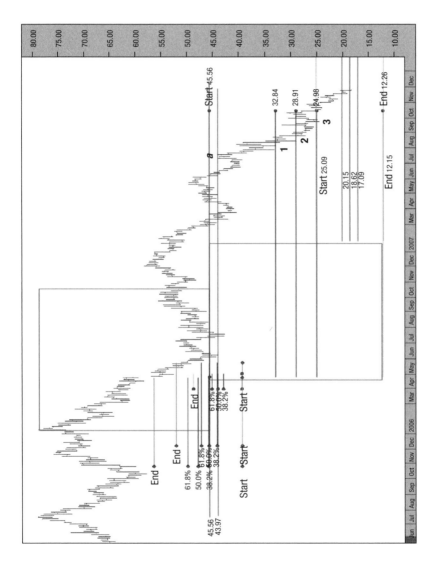

FIGURE 3.1   Centex 3-Day Chart

Connie Brown, www.aeroinvest.com. Source: Copyright © 2008 Market Analyst Software

the range from the higher confluence zone at $45.56 to the market high in this chart. A box is used just to measure the range. So the width of the box has no meaning in this application, just its height. I draw a second box over the first and drag the second box, being careful not to change its dimensions, to a position that starts at $45.56 and mirrors the first range. The range of the box ends at $12.26. That is a target too far away from the entry level at $43.97. (The price is on the left of the zone's horizontal line.) As I know I can add, subtract, divide, and multiply ratios, I elected to subdivide the new box with a Fibonacci retracement tool to determine the 38.2, 50.0, and 61.8 percent ratios within the box. A rational target in my comfort zone is the first level at $32.84. From this single projection, I do not know if my target will be minor or major support. I do not know what probability to give this target either. While it is a valid target, I have other ways to accomplish this task that experience has shown is better. This method is very useful for the Elliott Wave–challenged trader. There is no wave counting involved. Just a simple use of mirror geometry and you have a good start when your indicators confirm.

At price lows *1, 2,* and *3,* the market is respecting the zone created at $45.56 that was used to create the equality mirrored swing using the boxes. There is a reason this works and I'll defer that discussion until the next chapter. Our focus will remain on the methods used to create the price targets.

If the market reaches price low *3,* you may find a bounce that your indicators warn will lead to further losses. If you know the Elliott Wave Principle, it is easier because you would see an incomplete wave structure in the decline into the low at point *3.* Momentum oscillators would also warn a final bottom is not in place. Whatever method you use, there is a big spread from the 61.8 percent line that runs through price low *3* and the bottom of the box at $12.15. So subdivide the last range from the 61.8 percent ratio at $24.98 and $12.15. (You will find $25.09 in the chart, as the tools used were deliberately spaced different so you could see them. This is also true for $12.15 and $12.26 that end ranges.) This decline is incomplete but a bounce will form.

How do I know a bounce will develop? *My indicators are used only when the market reaches a target zone.* The indicators then give me permission to develop a trade strategy or warn when my exit plan should

be followed immediately. Part of the exit plan might be to unwind
a portion of the trade and add that portion plus x percent into the
bounce. By focusing on momentum indicators only at the zone, this
method filters out premature and false signals. I will look at oscillators
again in Chapter 5.

How do I know how high the bounce will go? I repeat the process all
over again by starting from a price low to calculate resistance levels and
take subsequent price bar highs that started a strong thrust down within
the decline.

But how can I be confident this is the price support level the market will respect for a significant rebound? From this simple method
alone, I cannot answer this latter question. For this reason, we need
to continue.

In the chart in **Figure 3.2**, you are going to use more of the information and observations we developed in Chapter 2. In Figure 3.2, you see the same box from the high to the zone at $45.56.
But flipping the box down to produce a mirrored target overlooked
two important facts about this data set. The first thing never considered was the gap near $58.18. If you page back to Figure 2.10,
you will see a specific measurement was taken because the gap was
in this chart. The other fact not considered was the price low, *A*,
in Figure 2.10. We talked about why that price low was used rather
than the key reversal swing down that followed shortly after this
move. We discussed how the probability, and hence our confidence
of being right, was confirmed when the market showed respect to
the results in hindsight within Figure 2.11. Plus we cannot overlook
an important point . . . we were right! When studying Figure 2.11,
we found the market failed after a corrective rally into the 50 percent retracement. All this information is evidence why this market
is declining. In the example for Figure 2.11, we had not discussed
confluence zones, but price low *B* in Figure 2.12 is a major factor in
what we are considering now.

In Figure 3.2, you see the confluence zone at $43.97 from the work
in Chapter 2 in Figure 2.13. The market is saying this price low *is one
of the key levels it is using to build future price swings*. We have to use it. The
upper box dimensions in Figure 3.2 are the same as Figure 3.1, but now
we are going to project the box down from the lower confluence zone

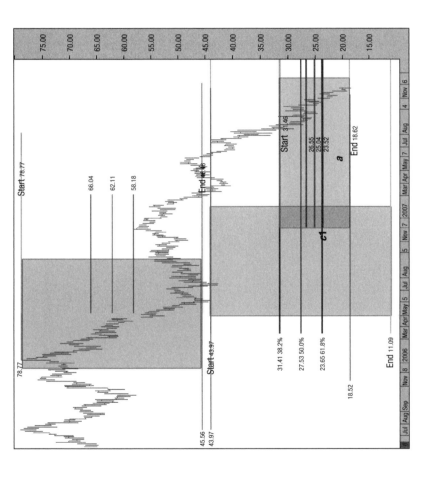

FIGURE 3.2 Centex 3-Day Chart

Connie Brown, www.aeroinvest.com. Source: Copyright © 2008 Market Analyst Software

at $43.97 that we know is a major marker for this market. The spacing between the two boxes, or confluence zones, is the same width as the gap. No surprise, as often markets use the gaps as measuring features elsewhere within a chart. By knowing we gave consideration to the gap and the key level at $43.97, we know we have built upon the earlier knowledge the market gave us. Our targets will be more accurate, though we need to define confluence zones to identify the difference between major and minor support targets.

In Figure 3.2, the market has fallen to a level at $18.52 that allows us to consider another proportional subdivision of the lower box. The lower box was subdivided into 38.2 percent, 50 percent, and 68.2 percent ratios, and the prices are at $31.41, $27.53, and $23.65 respectively. In Figure 3.2, we also created a box between the 38.2 percent line at $31.41 and the current price at $18.52. When this smaller box is subdivided, we discover a confluence zone where an overlap forms at 23.52–23.65. This confluence zone is marked *c1* and will prove to be important. All other lines in box *a* from $31.41 will be secondary. Why? The other Fibonacci ratios all stand alone.

The next chart in this series of 3-day Centex Corporation price projections is **Figure 3.3**. This chart shows several ranges subdivided, and additional confluence zones at $18.52 and $15.76. Draw a horizontal line at any confluence zone. The reason is found in **Figure 3.4**.

The reason you want to draw horizontal lines through the confluence zones, in some cases marking the highs and lows of the range itself, is you can delete all the other details from your screen. Most of the Fibonacci lines and boxes are deleted leaving a clean screen in Figure 3.4. Now other methods will use these confluence levels that we will discuss at another time.

At the start of Chapter 2, we reviewed the terms and methods to create ratios, mean values, and proportions. Some of the secondary subdivisions in Figure 3.3 use proportional measurements. The Fibonacci ratios subdividing a range are in every example, but the mean has not been used. In **Figure 3.5**, we need to stay with the same time horizon and stock to allow these next concepts to be clear. We are going to use two facts from our earlier work to develop a different way of making a price projection for Centex Corporation.

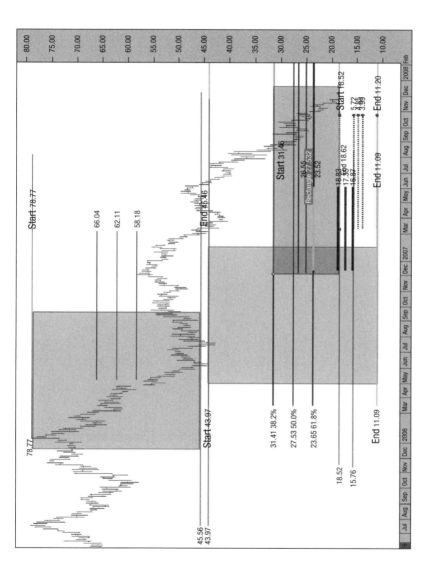

FIGURE 3.3   Centex 3-Day Chart

Connie Brown, www.aeroinvest.com. Source: Copyright © 2008 Market Analyst Software

49

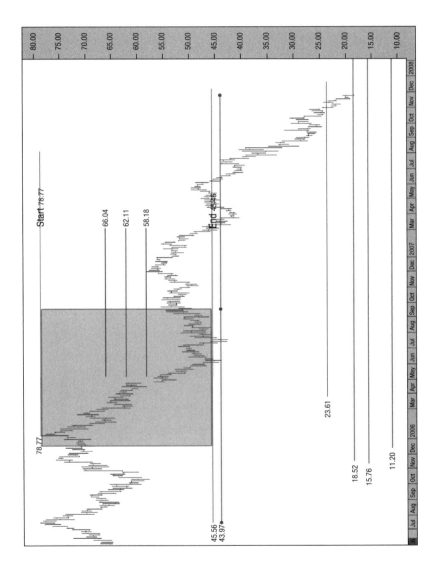

FIGURE 3.4   Centex 3-Day Chart

Connie Brown, www.aeroinvest.com. Source: Copyright © 2008 Market Analyst Software

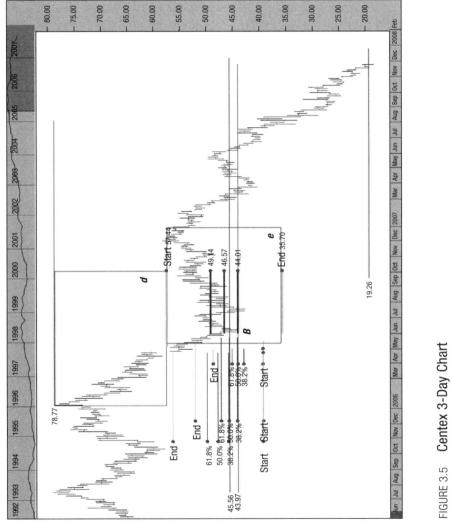

FIGURE 3.5 Centex 3-Day Chart

We know the gap near $58.18 is important. We also know gaps often form 50 percent subdivisional lines within the developing price geometry. You will use these facts now to your advantage. Define a box from the middle of the gap to the price high. This first box is marked *d* in Figure 3.5. That will become your arithmetic mean within a proportional equation that you need not write as a formula to understand.

Copy the box and create a mirrored placement under the first box. The new box in the middle of the chart is marked *e*. Within box *e*, you will subdivide the height of this box into the Fibonacci ratios. Take notice of the ratio that forms at price $44.01. All the old ranges that defined the two confluence zones in Chapter 2 are visible in this chart. The confluence zone at $43.97 that crosses the critical pivot at price low *B* has just been confirmed again. You are using different methods, and finding the price low at *B* and the midpoint of the gap continues to anchor key measurements. This is exactly what is meant when I said the market forms milestones that tell what key levels are being used to form future swings. All these calculation are derived from the market decline without benefit of the market data that formed in the rally. The reason for not showing the data in the rally is to learn how to do this when it is hard because new price lows are developing. The same methods apply when the market is breaking into historic new price highs. We will look at that scenario, as it is fitting for China, India, Australia, and many of the equity markets being dragged up by these indexes. But this 3-day Centex chart is building the skills you need to tackle the tough international scene we will discuss later.

Figure 3.6 is the third box in the series. Move a copy of the first or second box down to the bottom of the second box in the Centex Corporation data. You will again subdivide the range in the third and lowest box. Level *e*, or the 50 percent line, is respected by a small bounce up before the market works its way lower. Level *d* is also respected as resistance if you trace line *d* back to the left to pivot *d1*. Price level *c* shows how the market will use the Fibonacci ratios at *c2* as support, and *c1* as resistance. (In Figure 2.12, *c2* is level *B*.) The math would look ugly, but the geometric visuals are very easy to use and read. This is why Plato used geometry to explain the ratio that binds all things together and not the Fibonacci numbers that produce irrational numbers.

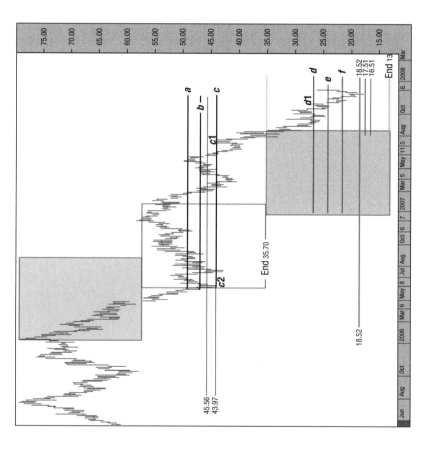

FIGURE 3.6   Centex 3-Day Chart

Connie Brown, www.aeroinvest.com. Source: Copyright © 2008 Market Analyst Software

53

An Introduction to Fibonacci
Expansion Price Targets

Another method widely used by swing traders is to select a range, as you see in **Figure 3.7**, and then project the 0.618, 0.50, and 1.618 relationships from a key price pivot. Figure 3.7 shows the 3-day Centex Corporation chart and the selected range from the high at point *A* and the now familiar price low at point *B*. Most people will use the price low to the right of point *B* where it appears to be the end of the swing. You are not going to use that lower pivot because of the important relationships we uncovered earlier in this chapter at *B*. The high of the corrective rally that follows into *X* is then used to start a Fibonacci expansion price projection. The 0.618, equality (1.0), and 1.618 proportional relationships are the industry standard ratios created from the range *AB*. Most books use diagonal lines to mark the swings being created; however, this is mathematically incorrect. You never give the slope or measurements on a diagonal axis any consideration. The projections are always parallel to the y-axis, as you see in Figure 3.7.

The equality, or 1.0 target, appears to be a confluence target, but it is really the same measurement as the one made in the last example in Figure 3.6 where the gap was used. The real confluence target is missing from Figure 3.7 because people generally only use 61.8, 1.00, and 1.618 for Fibonacci expansion projections. They often miss the confluence zone because the 1.382 and 1.500 relationships should have been included. Figure 3.7 introduces you to the concept, but you will soon make a more thorough study of how Fibonacci expansion targets are used to differentiate between major and minor target areas.

Now we need to study Centex's longer-horizon data in **Figure 3.8** that leads into the price highs. We will need to define additional confluence zones within this data. All the charts from Figure 2.8 to Figure 3.7 kept us from seeing this data so it was clear we couldn't be influenced by the historical data behind the most recent downtrend. As I created the charts, I deliberately kept myself from looking back as well. Therefore, I am looking at the long-horizon data myself for the first time in this analysis discussion. The first thing you know: to start a support calculation, you must start from a price high and drag down to a price low. The question is where do you start the high? There is a

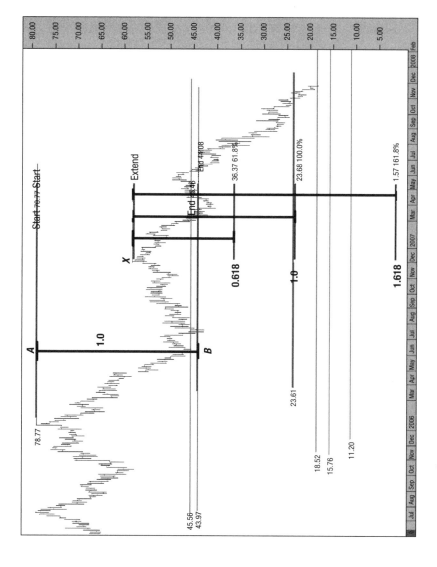

FIGURE 3.7   Centex 3-Day Chart—Creating Fibonacci Expansion Price Targets

Connie Brown, www.aeroinvest.com. Source: Copyright © 2008 Market Analyst Software

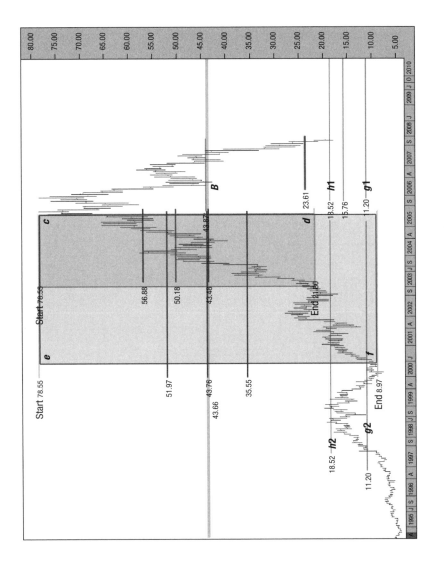

FIGURE 3.8   Centex 3-Week Chart—Longer-Horizon Data

Connie Brown, www.aeroinvest.com. Source: Copyright © 2008 Market Analyst Software

56

key reversal next to point *c. I normally truncate the starting key reversals for these range selections.* But I have to stop myself in this example. There is a double top and both tops produced key reversals. Strong directional signals should not be truncated. So this chart demands you start the range to be subdivided from the price high.

In Figure 3.8, the box corners of the selected ranges are marked. The first range is from the price high aligned near *c* to the lower price range at *d*. Why *d*? I find in my seminars everyone understands the concepts and then they get confused before the computer to do this. Nearly everyone has trained their eye to look at the price extremes that form the major swings, but they have never looked at the internal details of the data. So I know this is new for the majority of readers. We are looking to select a price low that starts a major price move. The price low that aligns with *d* is clearly the start of something strong, forceful, and relentless. The data that immediately leads into this explosion shows a period of higher highs and higher lows that form a back-and-fill coil. Take the low that finally breaks away from that preparation. These are the areas that let us adjust and stay in sync with the different contraction and expansion proportions developing within all markets. They are unique. *They require us to read the data.*

Behind the market high, I see lots of price lows with strong forceful moves within the rally, but you are trying to find the first levels of confluence that may be near price low *B*. The second range will again start from the high. Always begin from the same level. Then the price low for the second range stops at *f*. As I dragged my mouse down to look at the swings under the low of the first range at *d*, I found that the Fibonacci ratios were just noise within the chart. This is why you have to start from a price high to define support. If you start from the bottom, you have no other option than to pick the final high. If you stay with your vendor's thinking, you will always remain a beginner. I never complained when I managed a fund, as I knew most vendors locked the majority of traders out of using these methods. But later I learned from writing a report for investors and institutions, that you can advertise your exact price target and still be able to use it. Everyone has to tweak his or her results. It is amazing to me that exact price zones never seem to be messed up when large trading firms have been notified of their location.

In Chapter 5, we will study how to create price objectives when markets are moving into new market highs. You will see that the confluence zones we are developing within Figure 3.8 will be used to project future price targets for new market highs.

A smile came across my face when the second range was completed in Figure 3.8. The subdivided ranges *cd* and *ef* define a confluence zone right along the price level that intersects the much referenced price low *B*. No surprise to me as this will happen all the time. Use any time horizon, any Fibonacci projection method correctly, and the most significant milestones within the data set will always reappear. But if you never create multiple projections, you'll never know where they are hiding. You now know you were correct to use the price low at *B* and not the actual low that fell just to the right of it in all your prior calculations. This area is marked with a horizontal line that runs across the entire chart. It is also an exceptional chart to show why short-horizon traders must work with long-horizon charts as well and vice versa. There will come a day when the short-horizon trader's data will tell the trader point *B* is important. But just how important is unknown unless you work with this longer-horizon data as well.

One last comparison has been added for you in Figure 3.8. The confluence zones developed earlier at levels *h1* and *g1* have been extended to the left at *h2* and *g2*. I want to use this opportunity to reinforce an earlier comment that markets will show respect to these confluence zones in the future and in the past. In this chart, a major spike reversal developed just to the confluence zone *g2*. The spike reversal is just to the right of the label *g2*. Notice also the lengthy consolidation that developed along *h2* into the 1998 and 1999 highs.

Figure 3.9 shows Centex in a 6-month bar chart and becomes a rather dramatic clarification how confluence zones, derived from shorter time horizons that appear multiple times, may have long horizon implications for a market. Look at price low *B* and the corresponding breakdown three bars to the right with a down arrow. This zone was calculated from different approaches and entirely different internals, but each time we had a sense something important was forming at this price level. In this time horizon, it is clear the price zone at $43.48 to $43.87 was the most significant pivot point.

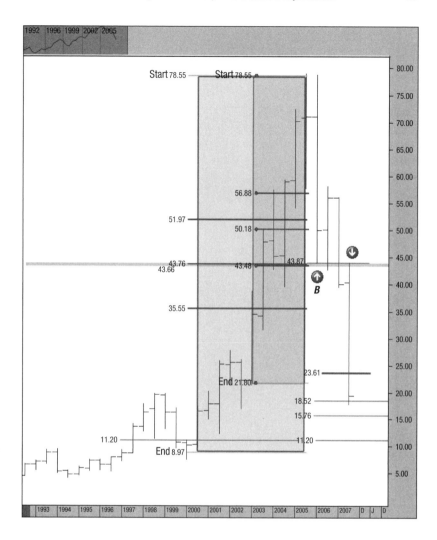

FIGURE 3.9   Centex 6-Month Chart—Short Time Horizons
with Long-Term Market Implications

Connie Brown, www.aeroinvest.com. Source: Copyright © 2008 Market Analyst Software

You are ready to focus next on rally examples and walk through
the steps needed to create price targets for markets making new market highs in Chapter 5. In Chapter 6, we will address some of the

problems that develop in different price character, such as contracting triangles, which warns when a market is rescaling. We will also cover adding technical indicators and other methods so you know what action to take when a market reaches a target zone. But before we move forward with more charts, this is a very good place to digress and explain why these methods work the way they do. It also goes back to an unanswered question you still may have about how to transfer the Fibonacci spiral into two-dimensional charts. Learning how to see the mysterious thread that connects the spiral galaxy (see **Figure 3.10**), the nautilus shell, your DNA, and great masterpieces in art and music, including architecture such as the Colosseum in Rome, follows next.

FIGURE 3.10 **Galaxy Messier 101, Hubble Image: NASA and ESA, February 28, 2006**

Acknowledgment: K.D. Kuntz (GSFC), F. Bresolin (University of Hawaii), J. Trauger (JPL), J. Mould (NOAO), and Y.-H. Chu (University of Illinois, Urbana)

CHAPTER 4

Bridging the Gap Between the Nautilus Shell and a Market Chart

As discussed in Chapter 1, Leonardo Fibonacci said in the introduction to his major book, *Liber Abaci*, that he accompanied his father Gulielmo on an extended commercial mission to Algeria with a group of Pisan merchants.[1] The Pisan merchants would have traveled through the well-known Arab Mediterranean trade routes. Of greatest interest to us are his travels through the Fertile Crescent, or an area known in 6000 BC to 400 BC as Mesopotamia (Greek for "between the rivers"). Mesopotamia was a very fertile flood plain between the Euphrates and the Tigris rivers, now Iraq, northern Syria, and part of southeastern Turkey, and extending to the Persian Gulf.[2] The area was frequently invaded, and texts were often written in many different languages. In ancient times, Babylon was a city of great fame. Although the biblical story of the Tower of Babylon is how most people in the West recall this city, few know that all mathematical texts we have from 2500 BC to 300 BC are Babylonian.[3]

The ancient Babylonians knew how to create the golden rectangle. Excavated by Hormuzd Rassam from Sippar in southern Iraq, the Tablet of Shamash (Babylonian, early ninth century BC) now has a permanent home in Room 55 in the British Museum. The tablet has a length of just 29.210 cm and a width of 17.780 cm, but it is of great historical importance. (See **Figure 4.2.**)

The Tablet of Shamash replicates the golden rectangle, and most people pass by this tablet in the museum without a glance. The depth of understanding that this tablet displays of the peoples of its time

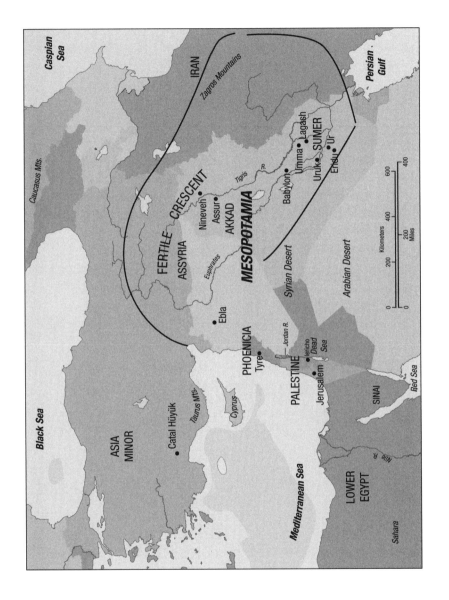

FIGURE 4.1   Map of Mesopotamia

62

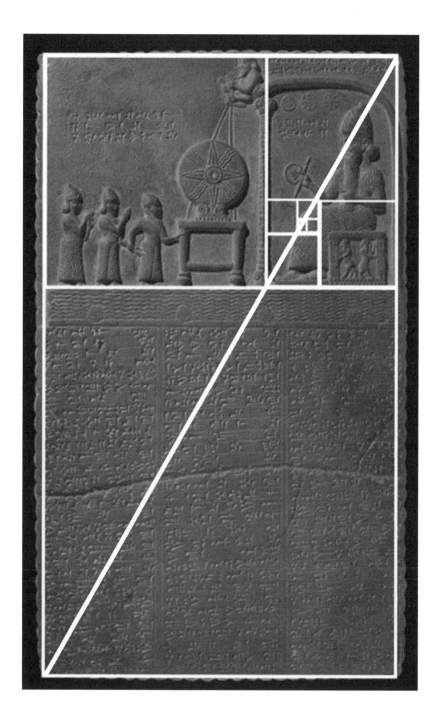

FIGURE 4.2   The Tablet of Shamash

is staggering. The stone tablet shows Shamash the Sun God seated under an awning and holding a rod and a ring, symbols of divine authority. Just to the left and above his head (under the arched rod) are the symbols of the Sun, the Moon, and Venus. On the left is the Babylonian king Nabu-apla-iddina between two interceding deities. The Chaldeans, who lived in Babylonia, developed astrology as early as 3000 BC, and the Chinese were practicing astrology by 2000 BC.

The tablet is interesting because of its distinct golden rectangle design, but of even greater interest are the three symbols of the Sun, the Moon, and Venus above Shamash's head. The mathematical relationship between the Sun, Venus, Earth, and its Moon form several phi relationships that one might not associate with the tablet if it did not clearly show the awareness of the golden sector.

An example of the golden sector is the diagram of the Sun, Venus, and Earth in **Figure 4.3**. I've drawn a 1 by 1 square and divided it in half with a vertical line. Now imagine this square on the ground, and then place a peg at point A to stretch a rope out to point B. Then stretch the rope to draw an arc on the ground from B to point C. Now move the peg over to point D and measure a rope from D to C. Next, stretch the rope with length DC in an arc to create point E. Points D, F, and C are the near relationships between the Sun, Venus, and Earth. But why did you draw point E? Point E[4] would produce what is known as the pyramid triangle, DEF, or the Great Pyramid of Khufu in Giza, Egypt. In ancient Egypt, rope stretching was a skill referenced and diagramed in several papyruses. You would have been known as a "rope stretcher" and highly respected.

Other relationships to Phi develop between Earth, Venus, the Moon, Mars, and Mercury. Only in modern times with the help of NASA could we determine that an average distance of all planets, with the largest asteroid Ceres, relative to Mercury in astrological units, equals 1.618.[5] There is no question from whom or where Leonardo Fibonacci gained his interest in Phi and phi. But even well before Leonardo's time, the ancient Babylonians did not use the Fibonacci numbers; they used geometry to produce the ratios phi and Phi. That is how we are going to use them as well.

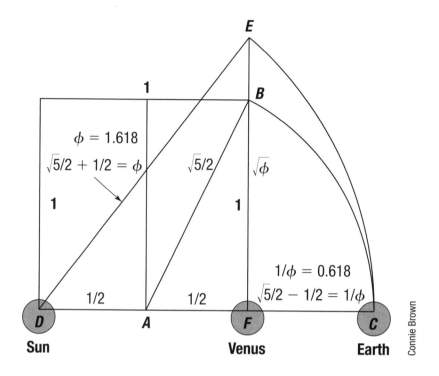

FIGURE 4.3   Phi Relationships

Source: Connie Brown, www.aeroinvest.com

So we know Phi (1.618) and phi (0.618) have been known and utilized for a long time. In order to show how to map the golden spiral, we need to learn new skills that will help bridge the gap between the mathematical model and the reality of having to work with price data that expands and contracts.

Proportional Analysis

The best place to begin is in the "pit." No, not the S&P pit, but the ancient pit of Rome: the Roman Colosseum. The Colosseum, with its "gladi-trader" connotation of do-or-die, offers an interesting way to introduce the new tools you need to bridge the gap between the proportions found in a nautilus shell and their reappearance in your financial data. More important, it will help explain why you want to identify the proportions within your charts from internal pivots and other key features such as gaps, truncated spikes, directional signals, and segments that start the larger meltdowns and melt-ups.

Fibonacci may have displaced Roman numerals with the Hindu-Arabic number system for conducting business and currency exchanges throughout Europe in the thirteenth century, but the Romans were clearly infatuated by Phi and phi long before. The photograph of the Roman Colosseum has been graphically delineated in **Figure 4.4** in order to analyze the basic proportions and elevations using a proportional divider.[6] There are so many Phi and phi proportions within the Colosseum relative to key architectural lines; we will use this image to learn how a proportional divider is used.

Using a Proportional Divider

The proportional divider allows us to make easy proportional measurements on our computer screens or paper without making any notations on the chart. It also allows us to consider how to study any image or printed chart in a book to define Phi and phi within the image. The proportional divider is a drafting tool with two arms held by an adjustable wheel in the center (see Appendix A). The first thing to do is set up the proportional divider by loosening the center wheel. We want the two arms to be in registration, which makes the points at the ends align precisely to one another. The two arms have a registration pin on the inside of the arm to ensure this task is done correctly. When the pin inside the arm fits a small notch, the points will be together. Next, hold the two arms and points together and slide the center wheel so that a registration line aligns with the number 10 marked on the "circles" arm. The illustration in **Figure 4.5** is a detailed image of a correctly set proportional divider. Some dividers will just be marked GS for golden section.

The key here is to ensure the center wheel is tightened enough to separate the arms, but not so loose that the arms slide from their registration setting at 10 or GS, as seen in Figure 4.5. Don't loosen the

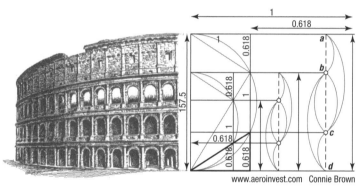

FIGURE 4.4   **The Colosseum in Rome**

Source: www.aeroinvest.com/Connie Brown

center dial so much that the parts all come apart. It takes very little movement of the wheel to accomplish this task. When the proportional divider is set in this manner, the spread between the arms will be different when the arms are opened. This is correct because anything measured with the longer side will produce a 0.618 relationship to this measurement on the short side. If you measure a length with the shorter side first, you will have a proportion of 1.618 to this measurement when you turn the tool over to use the longer side.

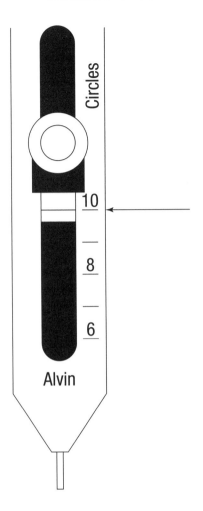

FIGURE 4.5　**Example of a Correctly Set Proportional Divider**

Source: www.aeroinvest.com/Connie Brown

Figure 4.6 illustrates the correct measurement of a data swing. Be sure your proportional divider has the registration line set at 10 (or GS). Separate the arms and use the ends that open farthest. Place the points in the circles marked *A* and *B*. This is how you would use the tool on the chart. You have to measure the y-axis difference *and not*

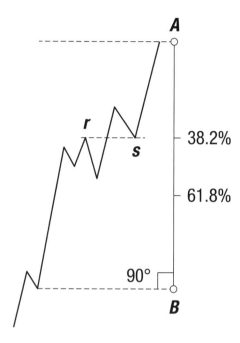

FIGURE 4.6   **Right-Angle Measurement for Proportional Dividers**

Source: Connie Brown, www.aeroinvest.com

the diagonal. Don't direct the points on the swing low on the far left and on the price high. Now that the long arms measure line *AB*, flip the arms around and never change the set arm spread from your first measurement. The shorter arms will fit exactly on the 38.2 percent line if you place one point on point *B*. If you start from point *A*, the second arm will fit on the 61.8 percent line. You have discovered why 38.2 and 61.8 are the same length.

In the analysis diagram of the Roman Colosseum (Figure 4.4), you will find two lines at the top of the drawn image. These lines are marked 1 and 0.618. You have to imagine points *A* and *B* now, but they are the start and end points of these lines. Check the registration of the arms again. Using your proportional divider, spread the points of the two arms apart and use the longest side of the drafting tool to measure line 1. Being careful not to move the arms, turn the tool over so you can see how the short side fits between the arrows on line 0.618 just under the line you first measured. You have just found your first proportional ratio of phi or 0.618 within the Roman Colosseum.

Like anything, a tool can be hard to handle at first. Setting the center wheel has a certain feel to it that you will soon have without thought. The first few times, it is essential you check the center registration line often. It is easy to make errors because the registration line shifts when you begin, because many do not set the center tension of the wheel tight enough. This is easy enough to fix if you just glance down on occasion.

Now you need to go back to the proportional analysis within the Roman Colosseum. Several 0.618 ratios have been marked, but you are going to find many more. Once this diagram is full of pinholes from your proportional divider, you will be a whiz with this tool. I have a beautiful art book of Leonardo da Vinci that is marked with small pinholes from the proportional divider. The book helped me learn how to use the tool and before I realized it my eye was trained to see this ratio with precision without the tool.

Now, this is where we depart from prior authors on this subject.

Introducing Rhythmic Wave Diagrams

To the right of the measured proportions of the Colosseum (Figure 4.4), you will find a *rhythmic wave diagram*. Rhythmic wave diagrams are being introduced to you now; I believe they are new to our industry, but they are not new to music theorists. This method of proportional analysis is a common way to define harmonic unity. In the chart of the Colosseum, two rhythmic wave diagrams intersect at the 0.382 and 0.618 proportional levels based on internal measurements of the Colosseum. These two wave diagrams are the easiest to relate to in the beginning. As an example, line *ad* is subdivided into 38.2 and 61.8 at points *b* and *c* respectively. In fact, they look similar to the results of a subdivided range into Fibonacci ratios. But the difference with rhythmic wave diagrams is their ability to show you the relationships between the proportional ratios. You will study rhythmic wave diagrams in Chapters 7 and 8 to begin to understand how the Fibonacci confluence zones begin to develop proportional ratios of greater significance. Not every confluence zone of support or resistance will be of equal importance. But some of the clustered zones of Fibonacci ratios will exhibit harmonic intervals in both the price and time axes. A rhythmic wave diagram is the first building block to a much deeper understanding of relationships between ratios, and it will help you trade with higher probability.

Harmonic analysis and rhythmic wave diagrams will go hand in hand. *The rhythmic wave diagram will paint a visual picture of unity between price swings to show that harmonic proportions are not serial.* As an example, consider points *a*, *b*, *c*, and *d* in the diagram of the Roman Colosseum. Arc *ac* skips over point *b*. The arc then feeds back into point *d*. What if all the confluence measurements in a chart fall on points *c* and *d*, and *b* is left standing alone? We would think price *b* was minor and prices *c* and *d* were major resistance or support targets. But in rhythmic wave analysis, what if point *b* becomes the start of a series of harmonic clusters with more ratios at the confluence points forming further ahead of point *d*? What would you think about price *b* then? How would you know to make a future price projection from point *b* *and not* level *c*? You developed the skills in Chapter 2 to prepare for this concept of confluence forming at major price targets. But you need to push this thought process much further as you see me infer from the questions I am asking now. Relationships between a grid of confluence zones impact your thinking about price projections, and they impact your method of defining time cycles. Time cycles are not serial. They plot one after another, but they are not related to one another in a linear simplistic series. You will see that harmonic and rhythmic wave charts can bring a far deeper sophistication to your analysis of price and time. *The reason the Fibonacci ratios that define confluence targets work so well within your charts is they identify harmonic proportions developing within your price data.* Harmonic proportions are not linear number series. We will develop these concepts in Chapters 7 and 8.

Geometry in the Nautilus Shell

The nautilus shell demonstrating the golden spiral is far more meaningful than the architectural proportions of the Roman Colosseum can reproduce. But the architectural analysis taught you how to use a proportional divider and gives you the freedom to study complex figures in books and other works of art. This is strongly recommended. A good place to start is in the artwork of Leonardo da Vinci, and then, study your charts in many time horizons. The purpose is to examine proportional relationships along the horizontal, diagonal, and vertical axes. In Chapter 5, we will apply these proportional relationships in charts. But it is essential to study more than just charts. The extra effort is fascinating and will help you train your eye to see proportional ratios more quickly and with surprising accuracy. Start with the horizontal as you did in the figure of the Roman Colosseum. With the basic skills and concepts you have in place now, you are ready to examine one of the most stunning forms in nature and then to see how the Fibonacci spiral applies to proportional unity within price swings.

The spiral of the nautilus shell holds within its geometry all the Platonic solids, the pentagon, pentagram, the Pythagorean 3-4-5 triangle, *and the geometry of the heavens*. The Pythagoreans used the pentagram as a sign of salutation among themselves. Its construction was a jealously guarded secret. The pentagon and pentagram are interesting because they are loaded with golden ratios. You will see in the golden spiral illustration (**Figure 4.7**), how the pentagram star forms from the nautilus shell's geometry. Few sacred geometry books describe *why* these forms are deemed so sacred. The pentagram, pentagon, and Star of David within the spiral are points whose angles mark a precise astronomical clock. They are measurements of time when specific planets come into orb or align to precise degrees of separation.

Everything in the heavens moves around everything else, and this dance to the *music of the spheres*[7] is mapped within the nautilus shell. As an example, every eight years Venus draws a perfect pentagram around Earth. The Moon squares the circle, and the inferior and superior retrogrades of the Sun and Mercury form the Star of David to geometric perfection. The pentagon formed the cornerstone of cosmological thought and represented the five wanderers starting with Mercury. Moving round the pentagon increases the planets' distance from the Sun. Wisdom keepers from the Rosicrucians of today tracing back to the ancient Pythagoreans, all look upon these forms of geometric perfection as proof there must be a divine creator. To understand how to map the nautilus shell onto a chart requires skills in

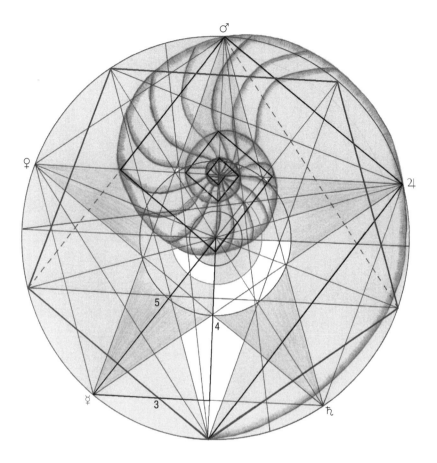

FIGURE 4.7   Golden Spiral Geometry

Source: Niece Lundgren and Connie Brown

geometry, harmonics, and astronomy. Without conscious awareness, you too have become a student of the ancient quadrivium. You will begin to develop these skills further in Chapters 7 and 8, and then apply them to your charts. This is the foundation of Gann analysis, and it is hard not to ask why this works so well in markets when you see a major price reversal where price confluence and time objectives merge. But I warn you now, when you begin to seek understanding behind the many whys, you will find a labyrinth of rabbit holes

FIGURE 4.8   **The Great Pyramid of Khufu**

Source: Private photograph collection of Connie Brown

---

leading to a path many have traveled before you. Together we are seeking the Truth behind the Fibonacci spiral and walk the steps of Plato, the Pythagoreans, and the ancient Egyptians. The pentagrams carved near the Queen's Chamber of the Great Pyramid of Khufu are shown in the photograph of Khufu's Pyramid in **Figure 4.8**.

CHAPTER NOTES

1. Cooke, 289.

2. The British Museum, Room 56: Mesopotamia 6000 BC–1500 BC. Objects on display in Room 56 illustrate economic success based on agriculture, the invention of writing, developments in technology and artistry, and other achievements of the Sumerians, Akkadians, and Babylonians, who lived in Mesopotamia at this time.

3. Cooke, 43.

4. Mad05, 3.

5. The relative mean distance of all planets and the largest asteroid Ceres is 1.618 when calculated in the following manner using astronomical units:

Planet	Mean Distance (in million kilometers) per NASA	Relative Mean Distance (where Mercury = 1AU)
Mercury	57.91	1.00000
Venus	108.21	1.86859
Earth	149.60	1.38250
Mars	227.92	1.52353
Ceres	413.79	1.81552
Jupiter	778.57	1.88154
Saturn	1,433.53	1.84123
Uranus	2,872.46	2.00377
Neptune	4,495.06	1.56488
Pluto	5,869.66	1.30580
Total:		16.18736
Average:		1.61874
Phi:		1.61803
Degree of Variance:		(0.00043)

During a meeting with Dr. Okasha El-Daly he commented, "Mercury had particular significance to the ancient Egyptians."

6. Please see Appendix A for a photograph of the actual drafting tool. I favor the 7½-inch Proportional Divider made by Alvin Company, as it is lightweight aluminum. The 10-inch Proportional Divider is made from steel and very awkward due to its weight. The lightweight 7½-inch duraluminum product is ideal for our use. A company in the finance industry sells this drafting tool under the name *precession ratio compass*. In my experience, it is not long enough for charts on 8½-inch by 11-inch paper or books. It is also made from steel and too heavy. If you enter "Alvin Proportional Divider 450" into your search engine, several suppliers will be identified. Alvin is a German company and the product is made in South Korea.

7. Godw, Taylor, Hall, Kap, Lev, Pin, Stroh.

CHAPTER 5

Fibonacci Channels, Angles, and Cycles with Oscillators

In Chapter 4, we analyzed the Roman Colosseum to see proportional ratios within the horizontal, diagonal, and vertical axes. It is important that we think about all three axes when applying this proportional geometry to technical charts. **Figure 5.1** is an overlay between China's Shanghai Composite Index and the Australian All Ordinaries Index. As 2007 comes near to a close, it is clear that North American equity indexes and Europe have lagged far behind China and its benefactors of Australia, India, and South African stock markets to name a few. But in North America, few traders monitor global equity indexes daily or even weekly, though it is essential in trading today's markets. North American equity markets lag; they are impacted by these global crosscurrents and cash flows.

In my library are many old and rare books. One small book by William Atherton DuPuy from a series called Factual Reading is simply titled *Money* (D.C. Heath and Company, 1927). It is rich with old photographs and clearly well researched, but of great interest now are pages 52 and 53. The nations of Europe were buying great quantities of war supplies during WWI from the United States, *and they were required to pay at least part in gold bullion.* By 1916, enough gold bullion bars had crossed the Atlantic that America housed $2 billion worth of it in this country. It was thought a huge sum for one country to house $2 billion in gold. But the war went on and nations were forced to borrow and buy more and more from America. The next year, America physically housed *one third* of the entire world's total gold reserves used

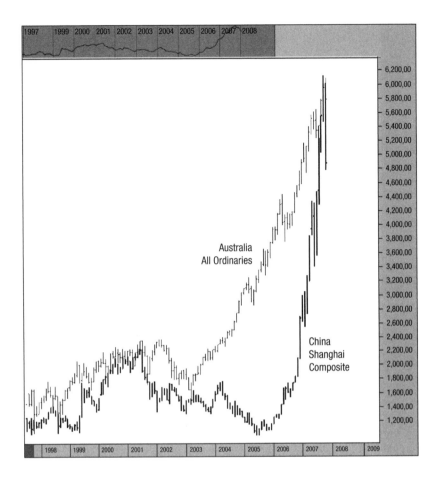

FIGURE 5.1   China Shanghai Composite and Australia All Ordinaries
Indexes

Connie Brown, www.aeroinvest.com. Source: Copyright © 2008 Market Analyst Software

for monetary purposes. The $3 billion in gold bullion continued to grow as the nations of Europe turned to America to rebuild after WWI. By 1923 the nation housed $4 billion of the metal in its coffers. The physical gold bullion peak came in 1924, but as we know, the country's equity bubble and the Roaring Twenties continued into 1929. People have long forgotten why the Roaring Twenties occurred.

Today, America is a debtor's nation and China has cash pouring into its coffers. The equity explosion in the China indexes is being fueled by this rapid influx of cash. History is repeating itself again on a different continent; the outcome will be no different. But technically, we can determine the levels of risk when fundamentals have historically never been able to time the end of a runaway bull market. In this chapter, we will cover how to forecast major market pivots in such an environment.

The first step is to recognize there is always a market that leads in the global arena. As an example, Japanese Government Bonds (JGBs) have led U.S. Treasuries for years. As JGBs moved so too did U.S. Treasuries move, weeks and sometimes months later. In Figure 5.1, the monthly chart showing an overlay of China's Shanghai Composite (bold bars) and Australia's All Ordinaries Index, you can see Australia bottomed in 2003 while China's market bottomed two years later in 2005. As you study the corrective swings between these two markets, you find Australia has been ahead of the moves in the China index. As a result, the momentum indicators you add, regardless of the formulas selected, will show a more mature pattern in the Australian charts than in the China equity indexes.

Some will look at this overlay chart and ask the question, would I use logarithmic scaling on the y-axis for such parabolic moves? The answer is no. In Chapter 8, you will learn the Fibonacci confluence zones you are uncovering now are subsets within a harmonic proportional series. Harmonic ratios are derived from prime numbers raised to an exponential number and therefore they become logarithmic. For that reason, I do not want to ever use a log scale within my price axes distorting the raw data. This will not be clear to all readers at this point, but some with a background in this area may have found it difficult to move forward if the question was not at least acknowledged at this point. Not to worry, as the foundation needed for all readers new to the concept of harmonics will be discussed in Chapters 7 and 8.

Never trade an index alone, as there is always a market that is slightly ahead and correlated to the market of greatest interest to you. You might not ever trade the leading or correlated market but if it is leading, you can use it as a bellwether indicator. In the book *Breakthroughs in Technical Analysis* (Bloomberg Press, 2007), you will see I used the 3-month Eurodollar in an inverse relationship to crude oil, as the 3-month Eurodollar is leading oil. Expect the inverse relationship to uncouple as the parabolic rise in oil matures. You will then lean on a different market. Consider stocks within a sector, similar bond maturities within Government yield charts as a global comparison, global equity indexes compared, but never analyze a single market alone. In this chapter, we will detail how to project a market bubble using Fibonacci geometry. Our goal is not to define the final top of a parabolic market; we need only find the significant pivots of resistance that a market will give a high degree of respect toward. Once we are short, we will be able to see from our indicators when the market will not hold major zones of support after a rebound from our targets. As Australia is leading China's Shanghai Composite, and China is fueling the global rally, we will focus on the Australian SP/ASX 200 market.

In **Figure 5.2**, the Australian SP/ASX 200 is charted in a weekly time horizon. As we want to identify support to begin your task, we must start from a price high and work down to multiple price lows. We need to give greater care now when selecting price levels for defining a range, as *most spike reversals will never be used to start a range*. We will not truncate double-top directional signals, but key reversals or railway tracks rarely display an internal price close higher than a previous bar near the highs. That's the nature of a key reversal. Tight bar formations that develop head-and-shoulder patterns are often truncated so that the start of the range bisects the matching shoulder pattern. As this chapter unfolds, be aware that the internal subtleties are of greater significance than the blatant swing highs and lows in many cases. Why use data where an emotional trap has developed? The market will mathematically ignore the few trapped in a false breakout as well, and this can be proven.

Figure 5.2 shows the first range selected in the weekly chart of the Australian SP/ASX 200 Index. It is then subdivided into the ratios 38.2 percent, 50 percent, and 61.8 percent, as was done in Chapter 2. The range begins from a price that is not the final high. The internals within the chart will explain in a moment why the actual high was not used. The low of the first range is an internal low that begins a strong market move up. Always look for the bottom of these strongest

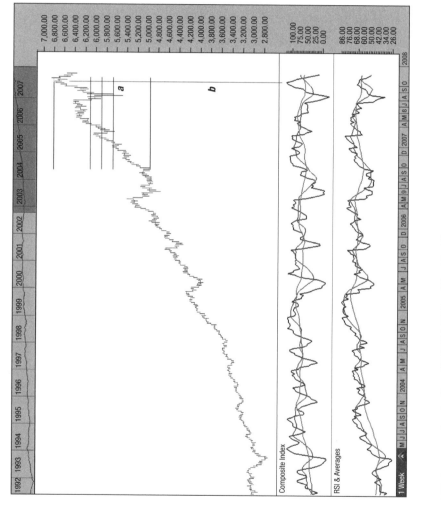

FIGURE 5.2 Australian SP/ASX 200—Weekly Chart

Connie Brown, www.aeroinvest.com. Source: Copyright © 2008 Market Analyst Software

bars to end a range. Conversely, when you are developing resistance levels, the highs of the strong price bars are used to end the ranges selected. (We used gaps in Chapter 2 as well. Gaps are always significant points to end a selected range for subdivision or to mark a mid-point within a swing.) It is better to have too many ranges than too few at first. But eventually you will save yourself the extra work and just select the bars that start strong moves. If you know the Elliott Wave Principle, use the start of third waves and wave iii of 3. If you are unfamiliar with this methodology, *I have an objective way to count waves* within this chapter you will like a lot.

Meaningful ranges will guide you by showing the subdivisions *within the range* have been respected before by the market. Take the time to look because you want to select ranges that help you account for underlying market expansion and contraction changes within the price data. If you look for the strong bars within a move, you are using the same milestones the market is using to create future highs in this case. The price bar at point *a* rests on a 61.8 percent retracement. If you had started the range at the exact price high, the internals within this range would not have shown any respect to the Fibonacci grid that resulted. I cannot emphasize the importance of this next point strongly enough. *Do not select a range to make the look-back price internals fit the resulting Fibonacci grid; select the range first based on the market pivot that fits the criteria that is the strong start of a move.*

The Composite Index in Figure 5.2 is an oscillator for which I released the formula in the book *Breakthroughs in Technical Analysis* (Bloomberg Press, 2007). It is used to identify divergence failures in the Relative Strength Index (RSI). The formula breaks the normalization character of RSI that forces the oscillator to travel between zero and 100. In this chart, you see the Composite Index has topped at the same price bars I used to start the range. Line *b* will guide your eye to the points being compared. Every additional range that follows will use the exact same start as the first range that ends at '*a*'.

Moving along a little faster than the examples in Chapter 2, the weekly Australian SP/ASX 200 now has three ranges defined. The first range is from near the 6,823 high to level *a*. The second range is from 6,823 to level *b*. The third is from 6,823 to level *c*. Both *b* and *c* start upswings within the rally. I do take into consideration the spread between the endings of these zones. If it is very close, don't bother. Remember confluence is made from different Fibonacci ratios. If the ranges selected are similar, they don't give any new information. As the three ranges are subdivided, a confluence zone is revealed near 5,691. If you look closely, the bar *after* (that was marked as point *a* in Figure 5.2) the capitulation key reversal held this confluence zone. Therefore, this area within this market is extremely important and the key reversal price bar into the actual swing low is not of interest. You will find a void of Fibonacci ratios under this confluence zone explaining why the market ripped down and then back upwards to the next zone forming under 5,958. Remember, confluence zones are formed when different Fibonacci ratios cluster together. The zone near 5,691 comprises a 50 percent and 61.8 percent retracement obtained from different ranges. The price level at 5,958 is not the cap of a zone, only a level of minor resistance. The zone itself falls just under the 5,958 labeled level. Market character between zones will be discussed in more detail in Chapter 6.

Oscillators can be useful as the price lows selected for range endings often align with indicators holding their moving averages or testing crossover formations within the averages on their respective oscillators. In **Figure 5.3**, a comparison should be made between the price lows and oscillator positions aligned with lines *a1*, *b1*, and *c1*. The oscillator position at *a2*, relative to the low at level *a*, has added significance because the Composite Index is warning that the RSI is failing to detect a market reversal when RSI fails to diverge with prices. The Composite Index then pulls back to test where its simple moving averages are crossing upwards to become positive at point *a3*. These subtleties are tremendously important, as they are the guides allowing you to adjust for any market expansion or contraction changes developing within your data.

A fourth range is added to the weekly Australian SP/ASX 200 chart in **Figure 5.4**. The new range has a new 38.2 percent retracement that falls within the same confluence zone near 5,691 (*m*). This repeating overlap of different Fibonacci ratios in the same area of the chart confirms this area within the chart is major support and is of great importance to this market. *All future price swing projections will be*

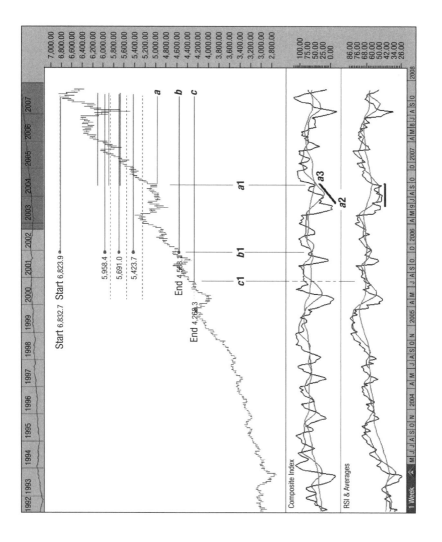

FIGURE 5.3   Australian SP/ASX 200—Weekly Chart

Connie Brown, www.aeroinvest.com. Source: Copyright © 2008 Market Analyst Software

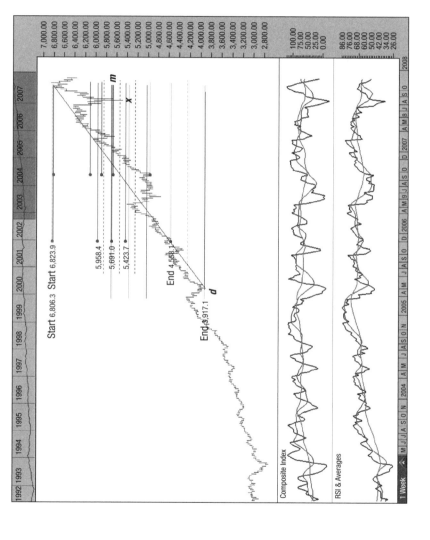

FIGURE 5.4   Australian SP/ASX 200—Weekly Chart

Connie Brown, www.aeroinvest.com. Source: Copyright © 2008 Market Analyst Software

*made using this hidden confluence zone that is well above the actual price low
that occurs at price low* x.

We can confirm that the market is using the confluence zone
at *m* to develop future price swings in the following manner. In
Figure 5.5, a box is used to measure the range between zones *ap* and
bm. The box is empty and has no significance to its width, only its
height, *ab*. I use boxes because they are easy to see. Next, draw a
box with the same height directly over the first. Then, drag the new
box up so its bottom sits on the top of the first box. The new box is
shaded a light gray, and the height at *cd* is the exact market top
before the hard break to the right of *d*. The market clearly shows
respect to this geometric measurement we have made, and it would
have been off if we had used the price high. This is how you know
to truncate the spike to start all the ranges. Eventually, you will see
everything aligns but is off the same amount as the spike you did
not truncate. Over time and with experience, you will know right
away where to start.

This is the smallest box I would use in this weekly chart. It is of
interest that price low *a* and price high *d* show the Composite Index
oscillator using its moving averages (simple 13 and 33 period averages)
as support into *a1* and resistance at the same amplitude level into *d1*.
The RSI is not as clear.

As the market respected the first box projected from confluence
zone *m*, you need to use a larger swing because you have lots of price
data under the range of the first box drawn. The next box is drawn
from the end of one of the selected ranges up to zone *fm*. No matter
what price swing low you start from, you will end the box at the confluence zone defined in the first steps. That is why you have to know
support in order to define future levels of resistance. At this stage, you
are still developing a geometric grid within existing price data. When
I draw a second box with height *ef* and project it from confluence zone
fm, I discover the top of the new box at point *h* has again been respected
by this market as a top forms. This is so major a find that you can view
the confluence zone at *m* as the midpoint of this market's developing
rally until it is broken by a larger degree decline. I am working from
a real-time chart and will make the next price projection, which will
be into the future for this market.

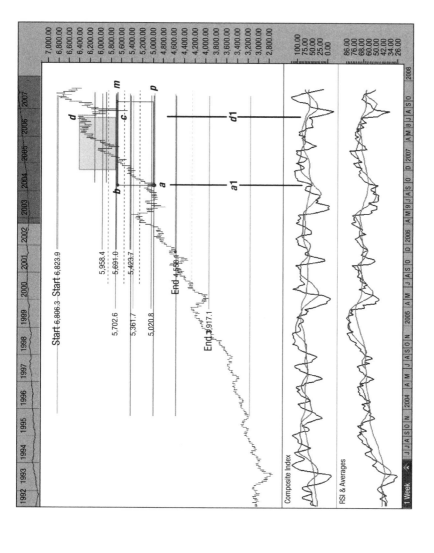

FIGURE 5.5 Australian SP/ASX 200—Future Price Swings Using a Proportional Projection

Connie Brown, www.aeroinvest.com. Source: Copyright © 2008 Market Analyst Software

87

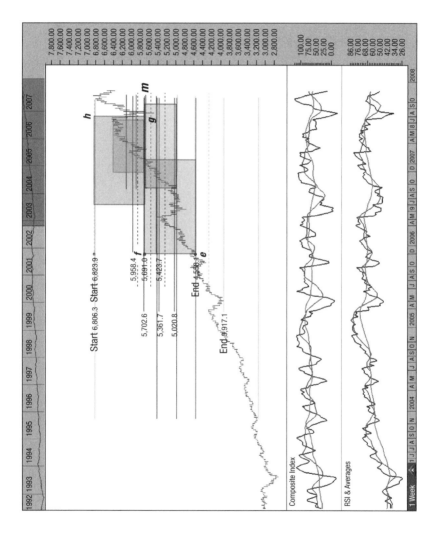

FIGURE 5.6   Australian SP/ASX 200—Finding the Significant Confluence Zone

Connie Brown, www.aeroinvest.com. Source: Copyright © 2008 Market Analyst Software

88

Figure 5.7 remains the weekly chart for the Australian SP/ASX 200 Index. The first steps to creating a future price projection have been completed by building a confident view the confluence zone at *m* is significant. Now, you draw an empty box the height *jk*. You used point *j* to define the ending price level for one of the larger ranges we needed to subdivide earlier. Now, you will use it again to take a measurement to the confluence zone. Move the box up to confluence zone, *m*. The top of the new box projects a market swing price level into the future. The price is 7,478. It is a single target and not confluence by itself. As the market topped near the 6,801 price level, 7,478 is a good distance away. So subdivide the box, *mn*, using a Fibonacci retracement tool. When this is done, you will find the market high is currently reacting to the exact 38.2 percent retracement within the new box, *mn*. This happens all the time that a future price target like *n* produces retracements showing respect to the existing data. It is how you can project more than one future swing. In this market, it is expanding, and you used the zone *m* to project and create all proportional measurements. If you are working with a contracting market, the zone that will prove key for all your measurements will fall short of an actual price swing. One fast way to find this area is to look for the strongest part of the move. Then when you create your confluence zones, you use the confluence zone that falls closest to the middle of the strongest swing.

If you know the Elliott Wave Principle, use third waves to your advantage. You will have an added advantage if you look for a fourth wave that falls into the vicinity of a previous fourth wave in an extending move. Mirroring fourth waves often give the midpoint away within these extending moves. You know you can add, subtract, multiply, and divide Fibonacci ratios to identify additional Fibonacci ratios, but people do not know how to apply this knowledge. Now you are finding this market is developing a proportional grid that is undeniably mathematical in its construction. *Trade the markets that form such grids and avoid the markets trapped in mathematical noise.* This grid work of confluence zones behind the price data is always present regardless of the market or time horizon.

Some markets are very thin and look like Swiss cheese with holes throughout the data such as esoteric currency crosses. Markets with

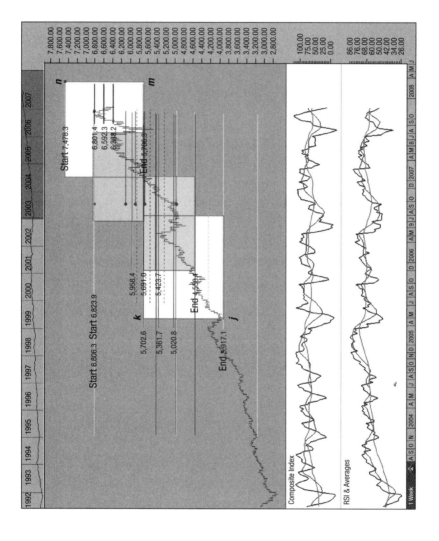

FIGURE 5.7   Australian SP/ASX 200—Creating a Future Price Projection

Connie Brown, www.aeroinvest.com. Source: Copyright © 2008 Market Analyst Software

90

this style of data can be converted into line charts first. Then use the line chart to calculate the confluence zones. If your market display preference is candlesticks, use the real bodies only to start and end a price range to create the grid structure. The real bodies in candles are very useful, but their limitation is there is too little data to view at one time along the x-axis. So keep this in mind and do not use large candles, but compress them as much as possible. Point-and-figure charts offer many visual cues to create a support and resistance grid using Fibonacci ratios. Truncate the isolated reversals and use the dominant blocks within the point-and-figure data. The confluence zones are accurate but the point-and-figure diagram has no x-axis for time analysis. Therefore, I prefer not to use this style of charting. Grids form in all price data character, *except triangles and termination wedges.* I will address triangles in the next chapter. It is not necessary to look for markets in stellar trends as you have the tools now to manage volatile market swings in any market.

In Figure 5.7, you can further subdivide the top box from *n* down to the first Fibonacci level at the 38.2 percent retracement near 6,801 to find other levels of minor resistance above the current market. I prefer to create additional boxes through the price data below *j* up to confluence zone *m* and then further subdivide them into simple Fibonacci ratios of 38.2 percent, 50 percent, and 61.8 percent. Soon you will find the confluence zone for future swings as the new boxes projected from *m* will be higher as the range is larger, but their proportional Fibonacci subdivisions *will soon overlap at future confluence zones.* My suspicions would be that a major confluence zone falls very close to the 7,478 level in this chart making it a major target within this developing rally.

We spent considerable time using Fibonacci analysis to project future price highs in a developing market rally. Most traders do not know how to project targets into the future except by Fibonacci expansion swings. When using a Fibonacci expansion tool, you need to change the default settings so proportional targets of 61.8, 100, 138.2, 150, and 161.8 percent are also determined. You will find the 138.2 and 150 ratios invaluable for revealing confluence zones. You will take a closer look at Fibonacci expansion swings in Chapter 6 and develop ways to account for expansion and contraction phases within a market.

The concept of confluence zones should be clearer to you now, and you should be able to see how markets develop mathematical grids within the price structure. When these skills come easily for you, and it does take practice and time, you will see the market confirming your efforts in any time horizon you elect to evaluate.

Once you are competent with the horizontal axis you can consider the diagonal and vertical axes within any chart. Don't jump to these axes immediately until the horizontal is very clear to you. It will be a mistake, as horizontal zones define the risk-to-reward ratios you need to control your capital drawdown exposure. The biggest problem I find with traders who attend my seminars is they take on too much all at once when they return home. Focus on one axis at a time and one new method at a time.

Many traders do not use the diagonal or vertical axes effectively. Traders often draw trend lines connecting price lows or highs, but much more can be accomplished.

One method is to create Fibonacci channels. **Figure 5.8** remains the Australian SP/ASX 200 index, but it has been changed to a shorter time interval of 3-days. I like a 3-day chart as it filters some of the noise in a daily chart and is faster than the majority who can only display weekly intervals. The reason you may see unusual display periods for creating my charts is the knowledge of knowing time ratios are significant. They help my indicators. I also use three moving averages with a fixed period interval because they provide a fast Gann target estimate. Therefore, I adjust time within my charts and not the period interval of my averages. If you can only see conventional time periods, analyze charts in a 4 : 1 ratio. In other words, a monthly chart against the weekly, a 60-minute chart against the 15-minute chart as another example. Entry and exit signals forming in these two time periods serve to filter out false signals in oscillators. The confluence zones in the larger time interval chart are major targets for the developing subsets within the shorter time period. *Use oscillators only when the market reaches confluence target zones.* Just these changes alone can dramatically change the probable outcome of a trade and your annual return.

In Figure 5.8, we created a parallel channel by looking at market pivots that test confluence zones on the horizontal axis. Also, all the Fibonacci grids have been removed, but not until horizontal lines

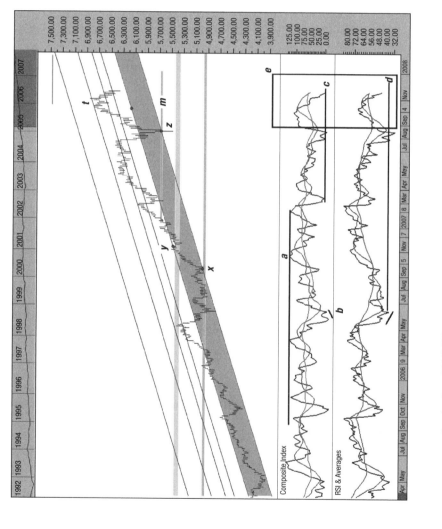

FIGURE 5.8   Australian SP/ASX 200—3-Day Chart

Connie Brown, www.aeroinvest.com. Source: Copyright © 2008 Market Analyst Software

93

were drawn across the screen to permanently record the price levels and confluence zone widths that were defined from earlier steps. The zones in Figure 5.8 have been further adjusted by using a transparency feature, so they appear as soft bands of gray running horizontally *in the background behind the data* and do not interfere with any analysis of the price data itself. There is no need of all the individual lines that form when using multiple Fibonacci retracements. Keep only the confluence zones themselves. The horizontal line on the top right of Figure 5.8 is the future price target that was developed from Figure 5.7. The zone most important for this market is at *m*. (Using a different color for this one zone on your computer may be helpful.)

Using points *x*, *y*, and *z* creates a channel. All three levels selected at points *x*, *y*, and *z* have shown respect to the confluence zones within the horizontal axis. It is very important to see that we are continuing to build from the zones defined in prior discussions.

Parallel channels are then projected in Fibonacci relationships relative to the width of the first channel defined. Each trend line above the channel has a Fibonacci relationship to the first range. In this chart, we see the recent price high has failed at one of the projected trend lines. In Figure 5.8, the Fibonacci ratios used in this chart are 61.8, 100 (equality to the width of the channel), 161.8, and 2.618 percent. The SP/ASX 200 failed at the 161.8 percent channel line (point *t*) relative to the channel width developed from the Fibonacci horizontal confluence zones. Like the Roman Colosseum diagram, this market is respecting Fibonacci ratios along two axes.

When the width of the first channel was being considered, the oscillators were examined. Point *y* has the Composite Index at a resistance level at point *a*. Point *z* not only tests major support, but also marks divergence between the Composite and RSI at a level that has been significant in the past. (Trace back the horizontal line drawn on the indicators.) Box *e* within the oscillators shows the indicators falling to the horizontal lines *c* and *d*. Another swing up will occur, and the danger point will be a swing pushing the oscillator to the underside of the averages crossing over towards a negative spread (within the box). The character of the indicator is significant and we should expect to see *W* patterns in the oscillator duplicate themselves at future times. This observation will help your timing.

Cycle Analysis

Time is without question the most important axis within our charts, but without price analysis skills, time analysis is useless.

The tools and methods to analyze time fall into three primary categories. The most common are fixed-interval cycles, in which traders try to find the best-fit interval to connect market lows. Because markets expand and contract in price and time axes, this method has limited success if you are looking for timing consistency that offers value for precise market entries and exits. Fixed-interval cycles of long and short periods will form their own confluence zones by creating cycle lows within a short period on the x-axis. Multiple cycle lows are an improvement, but even then, they are better used as an analysis tool, as the actual market timing to reverse in price can be significantly off for trading purposes.

A second method becoming more acceptable is to study astronomy and not fixed time cycles of any kind. In **Figure 5.9**, the 3-day Australian SP/ASX 200 chart shows vertical lines. As soon as someone sees astrological symbols, they assume this to be astrology. This has no relationship to personal natal chart reading or progression analysis. Astronomy is a science, and astrology is not. The ancient Babylonians studied the motions of the stars centuries before anything like a system of astrological interpretation developed. Astronomy is the scientific analysis of the planets, asteroids, and the Sun and Moon regarding position, angle, speed, retrograde movement, and much more. Astrological confluence to define high probable inflection points in markets works very well. If this is an area of interest to you, look into the analysis methods of W.D. Gann, who called this area of study the *Natural Laws of Vibration*. Gann had a marketing problem in his day, too, and never used the word *astronomy*.

The chart in Figure 5.9 is showing you how astronomy can help us time markets. In Figure 4.7, the last diagram in Chapter 4, the illustration showed how the sacred geometry figures form from within the nautilus shell spiral. The pentagram is loaded with Fibonacci ratios and many of the astronomical aspects that create the sacred geometry patterns contain the Fibonacci ratios of 0.618 and 1.618. It may seem unthinkable to you now, but NASA's website could become one of

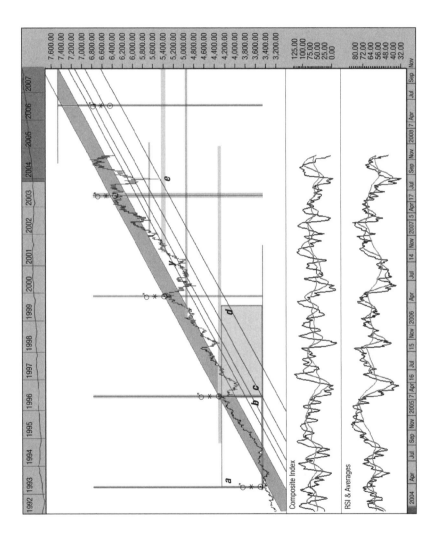

FIGURE 5.9 Australian SP/ASX 200—3-Day Chart

Connie Brown, www.aeroinvest.com. Source: Copyright © 2008 Market Analyst Software

your best friends for market timing. The vertical bars in Figure 5.9 mark when Mars and the Sun are within sixty degrees of separation and they are no more than +/−2 degrees apart. The width of the band will vary before they separate and exceed the +/−2 degrees criteria. When this condition of separation is realized by two degrees or less, we say the planets, in this case Mars and the Sun, are in *orb*. This cycle requires geocentric mapping where the Sun is viewed as an orbiting body around the Earth at the center. Even in this chart, it is easy to demonstrate the cycle is not a fixed-period interval between each aspect hit. Box *ab*, having to mark the diagonals of the box so the price data would not be hidden, was copied and moved to create box *cd*. This allows your eye to see that the next cycle is not of equal spacing to the first. Some cycle intervals are shorter and some have even longer spacing than box *cd*. This is the character of a natural cycle because of the elliptic orbital path of the planets around Earth. Cycles are not symmetrical and this method of analysis is worth the effort to learn.

Before we move away from Figure 5.9 to look at Fibonacci time cycles, study the Fibonacci diagonal channels in this chart. The upper gray channel is the original diagonal defined, connecting the tops at astrological time targets. The width of this channel forms the basis for projecting the Fibonacci parallel lines under the market. You will find price low *e* bottomed on the 261.8 diagonal ratio. As you scan the chart from right to left, observe the places where price respects the confluence zone created in the horizontal axis and bisects a Fibonacci channel in the diagonal. On several occasions, the intersection offers a good entry, but stops must be outside of confluence zones and never within the bandwidth. Long-horizon traders need to carry stops one zone away or possibly two depending on how the grid develops and the market's volatility. We will delve into this topic more in the next chapter.

We have considered fixed-period cycles and astronomical cycles that we know have Fibonacci relationships within the calendar intervals that create the sacred geometry forms from planetary aspects and retrograde cycles. Now, we will look at Fibonacci cycles along the horizontal axis.

The chart in **Figure 5.10** remains the 3-day data for the Australian SP/ASX 200. The vertical lines are Fibonacci time cycles. The spacing

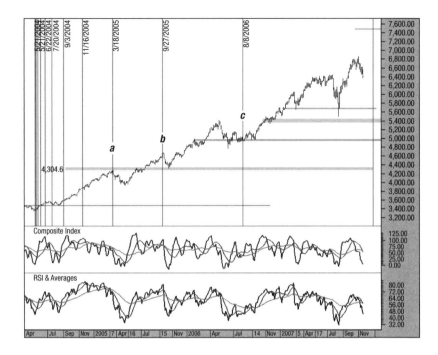

FIGURE 5.10   Australian SP/ASX 200—Fibonacci Time Cycles

Connie Brown, www.aeroinvest.com. Source: Copyright © 2008 Market Analyst Software

is a growth cycle that is 161.8 percent greater than the spread of the prior two cycles. The chart seems to have a few valid price reversals at points *a* and *b*. But the cycle interval that falls at point *c* is neither on the low of the downswing or on the start of the new rally. This is the weakness of using Fibonacci cycles as a trading tool. They are better viewed as an analysis tool than as a guide for timing trade entry and exit.

Many traders wonder if the global traders who have rapid access to these technical tools on computers actually reduce the effectiveness of these tools. In my library, I have a rare daily chart book that was in a private collection and printed in 1932 with permission of *Barron's* and the *Wall Street Journal*. Using the daily data from 1897 of the Dow Jones Rails and Industrials indexes, I know computers or easy-access communications by market participants does not contaminate the data. This is just an interesting example as I have made a more thorough study of all my technical tools and methods from the 1800s. (It is interesting to note a major freefall in the American equity market developed into 1903, similarly as the Nasdaq experienced in this twenty-first century nearly an exact 100 years later.)

The data in the Dow Jones 1897 Rails and Industrials chart (**Figure 5.11**) show a Fibonacci cycle added to this historical data. If I move the entire cycle series from the price low at *a* to a later start date when the rally begins at point *b*, the price high in Rails will align exactly. This is interesting as you do this when you define ranges for Fibonacci retracements. But the problem that results is the market does not respect any other pivots within the cycle's series. A modern-day technical tool applied to the 1897 Rails and Industrials indexes is no better than the chart we considered in Figure 5.10. Clearly, computers have made no impact on this cycle tool if you saw the conclusive results through the entire database. But even when confluence zones are developed using just the Fibonacci cycle tool, the results in my opinion are poor. But before discarding Fibonacci cycles as a tool to help with time analysis, you might consider a different application.

Conventional Fibonacci cycle tools are created from a single starting point. No one can use or read the first few projections as they are too close forming a band of noise. Consider Figures 5.8 and 5.9. Define a starting range between pivots and calculate the Fibonacci cycle as a series derived from the selected range. This is the same concept as described in Figures 5.8 and 5.9 for a diagonal axis projection. The method applied to the vertical axis is better than current Fibonacci time cycle convention.

If I were to ask you, "Name the year when the greatest number of banks and business failures occurred in the financial history of the United States," I bet you would answer, "a period within the Great Depression." It used to be my answer, too, until I encountered the following:

> . . . contraction of the currency began to be felt (as large quantities of currency left the country), multitudes of banks and individuals were broken. *The panic (that followed) caused the failure of nine-tenths of all the*

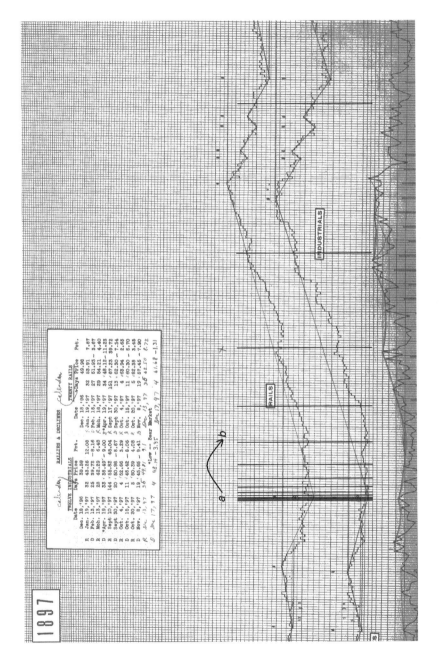

FIGURE 5.11 1897 Dow Jones Rails and Industrials Chart

Source: Private Publication by Robert Rhea that was printed by Permission of Dow, Jones & Co. Inc 1931

merchants in this country . . . Two-thirds of the real estate passed from the hands of the owners to their creditors.[1]

This is not a description of events during the Great Depression. It is the financial collapse that occurred in 1819 as a result of the War of 1812.

The Great Depression did not end in the United States until manufacturing had to ramp up production for World War II. But nearly one hundred years earlier than the Great Depression, in May 1837, a liquidity crisis caused the suspension of *all the banks* in the country. Samuel Benner's 1884 book called, *Benner's Prophecies of Future Ups and Downs in Prices*, is credited as the first book that provided market analysis and future cycle predictions. Benner states, "This year of reaction makes the second year in our panic cycles, and is eighteen years from 1819."[2] The panic peak, however, did not come until the Fibonacci cycle year of twenty-one years from 1819. Interestingly, the actual peak is a Fibonacci number. Benner's book discusses panic cycles for wheat, corn, cotton, hogs, railroad stocks, and iron primarily.

Think the year of 1819 is the worst year in American history? It was not. In 1883, an even greater number of business failures occurred, based on a total percentage of businesses. There is indeed a cycle under this boom–and–bust macro overview. Benner had collected government data from as far back as 1800 and covered economic cycles of prosperity and contraction for nearly one hundred years. Add this historical period in the United States to data from the 1900s and you have more than two hundred years to evaluate Fibonacci cycles within a macroscale environment. Robert Prechter duplicated Benner's cycle chart of past economic booms and busts from the 1800s by extending the exact graph of Fibonacci intervals into the 1900s to extend the cycle. It has had some market respect with the decline into the 1987 lows, but it has since added little timing value for traders. Benner's book forecasted a decline in 1891. He was right. His small book became widely read as additional cycle "busts" were foretold. But when the years arrived, the market failed to respect the targets. It would appear the macro results seem to be as fickle as seen in the shorter applications. Prechter, like Benner a century before, would find the macrocycle slip away.

Nevertheless, there is a reason for these seemingly fickle results. *Fibonacci time cycles are in fact only a subset of a cycle methodology that does not produce a linear series.* In other words, not every vertical line in this series should be given equal consideration, or in many cases, the cycle should not be given any consideration. It is not the tool at fault, but the application and understanding of its use. Even Benner in 1884 alluded to this fact. In the latter chapters of this book, we will revisit this discussion.

Fibonacci Angles

The final chart in **Figure 5.12** utilizes the 3-day Australian SP/ASX 200 to make an extremely important point. Speed lines, Fibonacci angles, and Gann angles all have greater value when projected from a price point intersecting a horizontal Fibonacci confluence zone. In Figure 5.12, a triangle, *abc*, is highlighted in gray. The bottom is set on a confluence

FIGURE 5.12 **Australian SP/ASX 200—3-Day Chart—Fibonacci Angles**

Connie Brown, www.aeroinvest.com. Source: Copyright © 2008 Market Analyst Software

zone. Point *a* is at the price that respects this confluence zone. You define the angle of the hypotenuse by drawing line *ac* and extending it so that it ends at the price highs that fell under the top of a box projection in Figure 5.5. By extending the hypotenuse line that defines the triangle, you create the most accurate angle possible. Fibonacci ratios on the horizontal are always points of interest defining this new technical projection on the diagonal. The range of side *bc* is now bisected into the 38.2, 50.0 and 61.8 ratios. Fibonacci angles result when the subdivisions along side *bc* are extended to point *a* and then forward in time. The middle-angled projection impressively stops the free fall that follows to price low *x*. Because you are working with proportional angles, a fourth has been added to show how a 161.8 percent projection would appear. It is marked Level 4 and bisects an extension of side *bc* and extends back to point *a*. The chart shows the market respected this Fibonacci angle as well at the high that forms to the left of *c*. A dotted line creates a rectangle by mirroring the triangle *abc*. It is of interest that the horizontal line across from *c* and the 161.8 percent angled line bisect at the price high to the left of *c*. As this is a confluence between the horizontal and diagonal axes, plus a time target based on the sextile aspect of Mars and the Sun on the vertical (see Figure 5.9), you have in essence just entered the analysis field of W.D. Gann. The tools are different, but the objective is not. When price, time, and geometric angles all come together, you have no greater point of confluence within a chart.

CHAPTER NOTES

1. Benner, 97–101

2. Ibid, 102.

CHAPTER **6**

Fibonacci Expansion Targets and Confluence in Time

THE CHART IN FIGURE 3.7 (Chapter 3) introduced the conventional method used by the financial industry for creating Fibonacci expansion targets. Convention says to take the range of a market price swing—for example, a price low to a high—and then project from the following correction low three Fibonacci ratios of the original measured swing. The standard ratios to project are 61.8 percent, 100 percent (equality to the measured move), and 161.8 percent. This method does not allow for market expansion-or-contraction price action within the larger trend. The projected expansion targets also yield a single price level that can neither be identified as major or minor resistance or support. Multiple ranges can be selected and the swings can then create confluence zones, but these still do not allow market expansion or contraction scaling within the data. This method can be less accurate than stringent risk management may allow. Therefore, this chapter will reinforce some of the earlier methods and concepts discussed and then advance the techniques further to improve the probability of conventional application.

The 10-Year Japanese Government Bond (JGB) has been a leading bellwether for the 10-Year U.S. Treasury Note for several years. In 1998 to 2000, this market developed a time-consuming corrective contracting triangle. Triangles are difficult, and this allows us the opportunity to take you step-by-step through a market that offers perhaps the toughest challenge this methodology experiences.

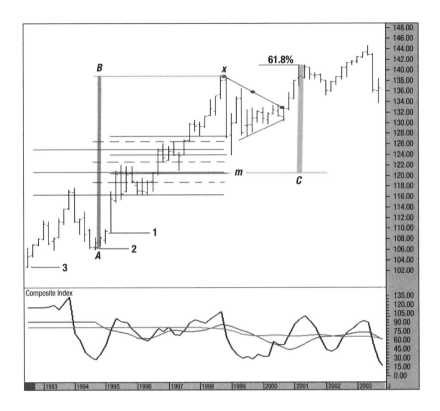

FIGURE 6.1   10-Year Japanese Government Bond—2-Month Chart

Connie Brown, www.aeroinvest.com. Source: Copyright © 2008 Market Analyst Software

The chart in **Figure 6.1** is a 2-month bar chart for the 10-Year JGB. In the middle of the chart, trend lines delineate where the triangle develops. We recognize this corrective pattern will lead to a thrust upwards, as it is a continuation pattern that will often resume towards the larger trend. The question is, where will the target thrust go above the triangle? The first step will be new to this industry. We are not going to use the dimensions within any part of the triangle to take a measured move and then project upwards from the resolution of the triangle. Why change traditional methods? Triangles develop five internal swings within the coiling continuation pattern. Elliott Wave analysts call the five internal swings waves *a*, *b*, *c*, *d*, and *e*.

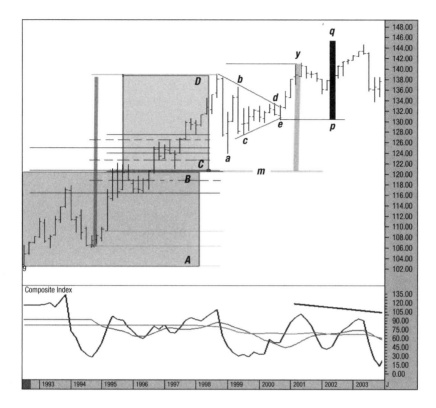

FIGURE 6.2   10-Year Japanese Government Bond—2-Month Chart

Connie Brown, www.aeroinvest.com. Source: Copyright © 2008 Market Analyst Software

(See **Figure 6.2.**) If traders use conventional methods, they must wait for the triangle to conclude before creating a target for the thrust measurement. This is poor risk management if you think about it carefully, as a trader cannot define the true risk-to-reward ratio before establishing a position. So we will project a target that can be identified well before the triangle has developed fully.

We will not define a measured swing as the first step but will identify where support occurred prior to the start of the correction. In Figure 6.1, we will find support levels created by subdividing ranges into Fibonacci ratios of 38.2 percent, 50.0 percent, and 61.8 percent. Three ranges were selected that start at *x* and end at the price lows marked *1*,

2, and *3* near the bottom left of the chart. As we drag our cursor down from *x*, we are looking for the strongest bars within the move, as we did in prior examples. *We are trying to find support under the triangle, so we know the confluence zone we need must be under the first swing down from* x. You will find other confluence zones in your data that fall within the triangle, but only one major confluence zone of support seems to form from within these three ranges that meets the criteria of confluence under the first leg down starting the triangle. The support zone is marked level *m* where a 50 percent retracement and a 61.8 percent retracement from two different ranges nearly overlap. Our real mission is not to find an old level of support far behind the correction. We are taking the time to walk through these steps so we can identify the midpoint of the rally that is developing. In this data, level *m* is defined as major confluence and support. This zone at '*m*' is going to be used to create the next target. The lowest swing within a corrective triangle is often higher than the mid-, or center, point of the developing rally.

Now we are ready to take a measured move of the rally before the triangle. The price low at *A* begins the sharpest leg up within this move. We discussed in prior chapters how a bar beginning a strong move is always an important internal milestone for any market regardless of the time interval we are viewing. (Using Elliott Wave terminology, we could say the start of a third wave or the third-of-a-third is important.) These ending points for ranges are always clearly defined, forceful price movement bars. In this case, the rally ends at point *x*. The range is marked by line *AB*. This is where this method deviates from tradition again. Now project a 61.8 percent ratio of range *AB* *upwards from the confluence zone at m*. The projected ratio relative to *AB* begins at point *C*, and the new price target is marked 61.8 percent. This is the first price high the market respected out of the triangle and follows the target with an eight-month pullback.

How can I be confident that the confluence support zone identified at level *m* is the midpoint of the developing rally? A proportional measurement from a lower price low to the confluence zone can be made to see how the results compare. Using the boxes again, I measure from the confluence zone *m* down to the start of my data in this chart. The height of the box is the measured range marked *AB*. A measured move of equal length *CD* is then projected upwards *from the confluence zone* m.

The market respected this equality target because the triangle begins from the same level as point *D* at the top of the second box. Now we know a mathematical grid line of great importance has been identified at *m*. The heavy black vertical line *pq* is the more conventional price projection the industry uses today, using a measured move from the triangle itself and projecting it from the end of the triangle. (Elliott Wave traders will reference the resolution of the triangle as wave *e*.) The method of convention is higher than the actual price high and gives no guidance for the first thrust out of the triangle that leads to the eight-month correction prior to the final rally. If your method varies to create a target out of a triangle, you may have used the length of wave *b* by extending the range back to the origin of the triangle (*x* in the prior chart). This target would have been off. If your method just took the range in wave *b* up within the triangle to define a target from point *e*, the measured move would be too high for the first swing out of the triangle and too low for the final top. By finding a support confluence zone behind the correction first, and then projecting a target, you will experience greater accuracy, and the target will be identified much earlier. This is one of the most complex scenarios you will ever encounter as triangles develop when the market is rescaling its proportional grid. In this case, the market was contracting, and that is why confluence was significantly below the first leg down in the corrective triangle pattern. It is understandable as the market was in the final stages preceding a five-year correction that is still developing.

In the shorter horizon, triangles will accurately move to a major confluence zone for the first price break or swing (in Elliott terms, wave *A*). It is then possible to develop an accurate target for the second swing, or wave *B*. But thereafter, the internals will break down through short-horizon price zones, and this character is of itself a clear warning you have entered a triangle and should not trade it. There is only one way to know a triangle will develop before it is on your computer screen: by finding a wide zone of astrological confluence targets along the vertical axis. Heavy congestion of astrological targets will cause this market action chop, or sideways coiling consolidation. The triangle usually ends right as the price data escapes the vertical congestion of time targets. This prior comment will likely make sense to Gann analysts only. But it is of such value to a trader, it is the primary

reason I began to explore the methods of W
know after a great run when I was going
brick wall of price chop and cause me to g
earlier gains back. Knowing when the marke
confluence zone along the vertical time axis ha
able addition. We will look at vertical conflue

One further thought about the boxes I use
offer a very distinct reference on the chart, bu
a chart program that does not allow them to c
move it around on your computer screen. If
use the Fibonacci expansion tool that your ve
ber to modify it. The Fibonacci expansion ta
the following ratios: 38.2 percent, 50.0 percen
cent (equality), 132.8 percent, 150 percent, and 168.1 percent. The
38.2 and 50.0 percent ratios will depend on the markets you trade and
the time horizon, but I never use the 261.8 percent ratio of the measured range. It is so far away, I would rather use the new price swings
that follow within the new data so a market starting to contract or
expand can be caught early in the process.

Just before moving on, notice the bearish divergence in the Composite Oscillator (bottom of Figure 6.2) compared to the final top and
the price high at y. This will reinforce earlier discussions to read oscillators only when the market reaches a confluence zone. Indicators tell
us how the market will react to the confluence zone, and if they give
permission to execute a trading strategy. By ignoring oscillators when
price is not at a confluence zone, we filter out the false signals. The
divergence of any momentum oscillator is insufficient information by
itself to establish a trading position.

In Figure 6.2, there is a near equality move using the conventional
method of measuring a range and then producing a new swing up
from the next pullback. The measured range is from the end of the
triangle at e up to the high marked y. Then projected as an equality
move from the corrective low to the right of y, it targets the market
top. Nevertheless, a confluence zone could have been used to project and identify this market top much earlier. The method will be
discussed shortly in Figure 6.4 once we develop additional confluence
zones.

The 2-month JGB chart is displayed again in **Figure 6.3**, but this time as a candlestick chart. This offers an opportunity to clarify why I often truncate market spikes. The top candle is called a *shooting star*. It is bearish as it denotes an exhaustion signal in the preceding uptrend. If you are unfamiliar with candlestick charts, the shooting star forms when the market trades higher than the market open, then the session closes below the open and near the lows of the session. The candle develops a real body between the open and closing prices of the chart interval. If the candle body is black, it means the closing price was below the opening price for the 2-month period. If the real body is clear or white, the close was higher than the opening price.

We are not going to read any traditional candlestick patterns from this chart. But we can use the clarity of the candlesticks to define support or resistance. In Figure 6.3, truncate the wick of the shooting star candle when starting the ranges to be subdivided into Fibonacci ratios. To define the end of the ranges, use the real bodies where the strongest moves develop. Three ranges were defined at points *1*, *2*, and *3*. The confluence zones clearly fall at *m* and *n*. A third zone grayed in will be used later.

The bottom of the third range at point *3* needs special mention. It is the longest real body within the developing rally. Ensure you include as one of the endings of your selected ranges a pivot that marks the longest single bar within the rally or decline whenever possible. The subdivisions within this range will develop the most important confluence zones from which to make future projections.

Figure 6.3 allows us to begin to consider market character between zones. Support zone at *n* is a band from 133.31 to 133.63. Remember, we are looking at a 2-month chart, and the confluence zone width is still fairly close together. The next zone at level *m* falls between 130.16 and 130.35. These confluence zones define precise levels of major support. The single level standing alone at 131.33 is viewed as minor support and not the upper collar of the confluence zone at *m*. Change minor support levels to a faint color or remove them entirely from your screen when the screen is congested. When the market falls to zone *n* just to the left of the *n* label, it recovers from this support zone and resumes the larger trend only to create a head-and-shoulders top. The breakdown through level *n* will fall rapidly to level *m* because no

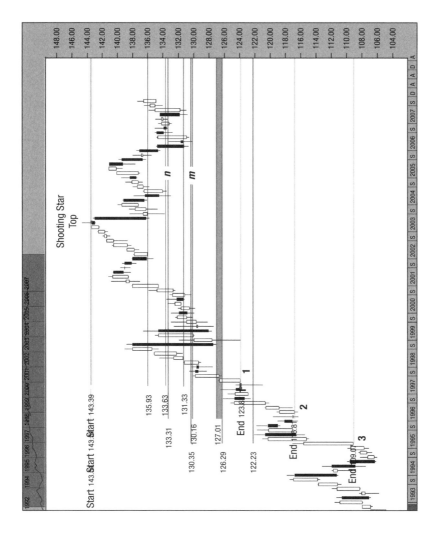

FIGURE 6.3 10-Year Japanese Government Bond—2-Month Candlestick Chart

Connie Brown, www.aeroinvest.com. Source: Copyright © 2008 Market Analyst Software

112

other Fibonacci levels are present to impede the decline. Expect this acceleration between zones to occur as the rule rather than the exception when there is a void, so to speak, between major zones.

For an Elliott Wave practitioner, this is how to know what pattern will develop *before it even begins*. Near the end of 2007, we saw in prior chapters, numerous stocks and global indexes formed parabolic rallies. Parabolic rallies *always* develop large voids between confluence zones defining support. These voids under the market create fast breaks or strong decisive swings as the market moves rapidly from one support zone to the next encountering little resistance. As a result, the manner in which the confluence zones fall under the market gives warning well in advance that the Elliott pattern known as a zigzag correction will develop. These three-wave corrections are big countertrend money producers, as the market character will be fast market conditions from one zone to the next. Often the middle consolidation in the correction, we call the *b* wave, becomes trapped within a nest of minor Fibonacci lines. When the second break comes out of the middle congestion, it is again a fast market condition through the empty void to the next major confluence zone.

Triangles form within a nest of congested Fibonacci levels, and the internal swings get caught between two or more major zones. Often there is no clean break between zones. Consider the triangle in Figure 6.3. Follow along the confluence zones at *n* and *m* to the left, and examine the triangle. There is a third zone with a spread of 126.29 to 127.01 just below *m* that is highlighted with a light gray color. The triangle pattern develops in stages along these three support zones. Triangles will be the toughest challenges we will encounter. The decline that develops from the market top clearly respects the support zones *n* and *m*. That's why I will trade corrective moves like the one we see to the right of the market high and try to stay clear of the final internal swings of a triangle. Other market price action to stay clear of are termination wedges, thinly traded stocks or data showing numerous long key reversals both up and down (market makers are letting stops run), and markets forming tightly congested chop with few well-defined confluence zones. Regardless of the time horizon in your chart, be it very long or very short, the market character or action between zones will respect the spread between the confluence zones created from ranges using strong moves or gaps. Gaps, it should be pointed out, are true market gaps and should not be confused with session breaks that occur because you are not plotting available night sessions.

It was suggested earlier that the market high for the 10-Year JGB market could be identified using a confluence zone. In **Figure 6.4**, a box from the confluence zone at *m* is used to start a measured range to the low of the price swing. This zone was ignored earlier because we had to find the first zone that formed under the triangle. Now this higher zone is used to create the measured range marked with a gray box. The height of the box is copied and shifted so that the bottom aligns with confluence zone *m*. Now subdivide the box into the ratios 38.2 percent, 50.0 percent, and 61.8 percent. The market high ends at the 61.8 percent target. You might have also observed that the first retracement up after the high has a candle whose real body fails at the 38.2 percent subdivision. Imagine the 50 percent line extended to the right and the next rally fails just under this level as well. The JGB 10-Year market is still working off this same price grid defined three years ago.

The next discussion will focus on the EUR/USD Forex market. This market will demonstrate a specific problem in price projections that can be solved, but my discussion will require a more advanced study of the use of Fibonacci numbers for time analysis. The chart in **Figure 6.5** is a 2-month EUR/USD currency chart. The extreme price lows and highs have been labeled points *A*, *B*, and *C*. A 61.8 percent ratio of range *AB* has been extended from price low *C*. This again illustrates the industry's convention for developing Fibonacci expansion ratios. The target is very precise, but three price bars representing a period of six months have freely passed back and forth around the target ratio. Many analysts may view this target of value, but traders need greater accuracy.

Figure 6.6 continues with the same EUR/USD market in a 2-month bar chart. This chart shows a more advanced confluence zone calculation. The difficulty is in knowing where to start the ranges that will be further subdivided into 38.2 percent, 50.0 percent, and 61.8 percent ratios. Study the price bars into points *x*, *y*, and *z*. The strong reversal at the actual price high between points *y* and *z* is a bull trap. Don't use bull or bear traps as a rule. Find where the market defined resistance prior to the trap. Point *x* is a strong bar moving up that ended the

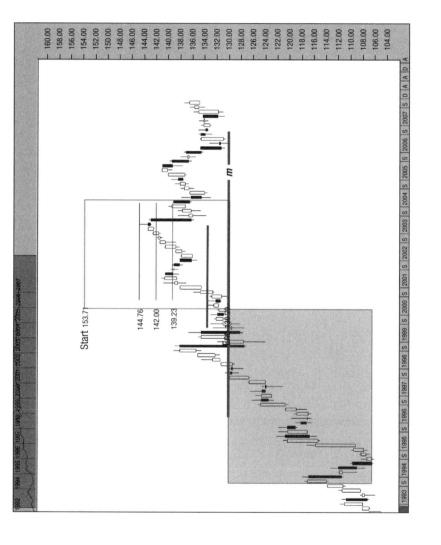

FIGURE 6.4　10-Year Japanese Government Bond—2-Month Chart

Connie Brown, www.aeroinvest.com. Source: Copyright © 2008 Market Analyst Software

115

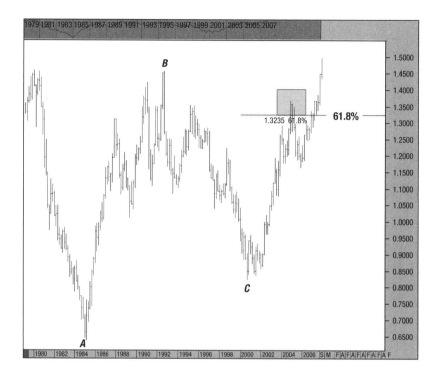

FIGURE 6.5   **EUR/USD—2-Month Chart**

Connie Brown, www.aeroinvest.com. Source: Copyright © 2008 Market Analyst Software

strongest momentum of the swing. The three bars, or six months, that followed made no substantial gain except to display market indecision. A sharp break then develops, and the strongest bar down begins from the same price level as *x*. A retracement rally is attempted with a massive failure into *y*. The market then produces the breakout failure and pullback that holds above confluence zone *m*. The retracement rally fails at point *z*. This next point is extremely important. Because points *x*, *y*, and *z* all develop at the same level *and the deepest retracement from the high between points* y *and* z *is respected by the market at* m, the key reversal false breakout should be truncated. The ranges selected all start from the *xyz* level and then end at strong price low bars within the rally. A confluence zone along price level *m* is identified.

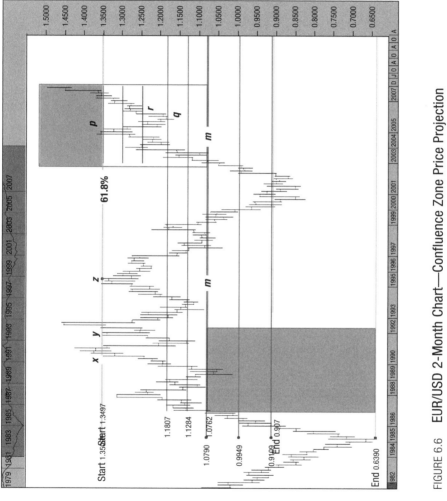

FIGURE 6.6 EUR/USD 2-Month Chart—Confluence Zone Price Projection

Connie Brown, www.aeroinvest.com. Source: Copyright © 2008 Market Analyst Software

The next step is to take a measured move from the confluence zone *m* down to the price low. It is a gray box on the bottom left. The range is then projected upwards from *m*. It is subdivided into the ratios 38.2 percent, 50.0 percent, and 61.8 percent. This method accurately warns that a failure into point *p* offered a good exit strategy for any long positions. The same level then facilitated a good short sale strategy at *p*. The bar to the left of *p* would be stopped out perhaps, so the entry at *p* is the second time, producing a solid move down without any position jeopardy. Always let indicators in a shorter time horizon confirm a longer horizon chart. The failure at target *p* would have divergence confirmation in a weekly and monthly chart, while the earlier pivot would not. Therefore, take profits into targets, but do not reverse positions without seeing the setup develop in more than one time frame. The grayed area above *p* is just to help visualize the larger trend's target, but indicators warn this target will not end the larger rally because divergence in two time horizons will not be present. Readers who want a more detailed discussion on the relationship of momentum displacement to amplitude range in multiple time horizons will find more in my book *Technical Analysis for the Trading Professional*.

The market did not stop exactly the first time it approached the target at *p*, but this level is a significant improvement over the method used in Figure 6.5. The question we are left with is, why was price pivot *p* so difficult to locate? There is an answer to this question. It was a horizontal level of resistance for price, but not at the vertical confluence zone defining a time target.

Fibonacci Applications for Time Analysis

Many readers will find the notations in **Figure 6.7** a new cycle technique. We do not need an extensive discussion about the field of study known as Gann analysis, in order to use some of the conceptual ideas of W.D. Gann within our Fibonacci analysis. One concept credited to Gann is an equality relationship between price and time. How Gann analysts apply this concept is to square the price axis to determine a target of equal units along the time axis. *Square does not mean price squared.* This is just a geometric equality relationship.

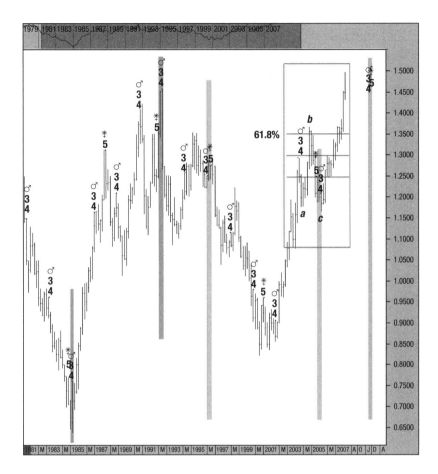

FIGURE 6.7   Advanced Application of Fibonacci Angles

Connie Brown, www.aeroinvest.com. Source: Copyright © 2008 Market Analyst Software

Fibonacci angles were introduced in the last chapter in Figure 5.12. The next step is to apply this concept of Fibonacci angles in a more sophisticated way. Figure 6.7 is the 2-month EUR/USD currency chart. There are five vertical lines through the chart showing confluence targets in time. One cycle is marked with the symbolic glyph for the planet Mars with the number 34 beneath it. This astronomical "hit" simply shows when Mars has moved thirty-four degrees along its elliptical orbit relative to earth. The second cycle is an asteroid being mapped every five degrees along its own orbital path as seen from the Sun. This is called *heliocentric*. A cycle progression as perceived from Earth is called *geocentric*. The two cycles will form major inflection points in time when they come close together creating a time confluence target on the x-axis. These two cycles form five vertical, or time, confluence zones. The EUR/USD market has respected each time confluence zone by forming a market reversal or pivot that resumes a strong resumption of the former trend. You will be able to identify the inflection points of high probability with some additional study, but don't assume the inflection point is a historic price top or bottom unless you become proficient in these methods of W.D. Gann. Most books on the subject discuss the methods of John Gann, his son, and they are far less accurate.

Some caution is needed about what you use to study these analysis techniques. Do take notice that neither of the astrological cycles displayed in Figure 6.7 has a fixed period interval between its cycle hits. Astrological orbit progressions, aspects, retrograde, ingress, speed, and much more can be statistically evaluated over large databases making them viable methods for various models. The notations could have been mapped under a price bar, or along the top or bottom of the page. *Place no meaning on the price level of these astrological notations.* They just have meaning on the vertical axis, and I have a preference to see them as close as possible to a price bar, because I want to know when a horizontal and vertical confluence zone come together. Like the confluence zones we created for price on the horizontal axis, the width of the bands on the vertical axis have meaning as well.

Now look inside the target box to the right side of the chart that was created for Figure 6.7. *Within this box, there is only one vertical time confluence zone, and the market is not trading near the 61.8 percent horizontal price target at that time.* The uptrend for the swing ended earlier within the target box at the 50 percent line when Mars marked a 34-degree progression.[1]

For traders well versed in Elliott Wave jargon, the pivot high within the box, marked ♂ 34 (Mars 34), is the *orthodox top* concluding a five-wave rally. A corrective pullback then leads to the high marked wave *b* within an expanded flat corrective pattern. *B* waves are ugly remnants of the former trend. In expanded flats they exceed the resolution of the impulse wave preceding them. From the 61.8 percent target in Figure 6.7, a wave *c* decline then unfolded to conclude the corrective pattern. An added measure of confidence occurs that a market reversal will develop because the wave structure of the correction ends at the same time as the confluence target in time. In addition, the price low of wave *c* was on a short horizon confluence zone not displayed within this chart. The rally favoring the Euro then resumed into the current data of December 2007. The confluence zone of greatest interest in Figure 6.7 is the one on the far right that is a future time target.

As stated earlier, astronomy is a science, and astrology is more of a mystical art. The financial industry has put a stigma on the science because of the perceived voodoo of the art and emotional presentations of financial astrologers targeting the mass retail market. But astronomy and astrology are not the same. If you wish to study astronomical cycles and need a place to start, begin with planetary retrograde periods. You will find a discussion in Chapter 5, "Price and Time," in *Breakthroughs in Technical Analysis* (Bloomberg Press, 2007). I was a contributing author and focused on the Oil market in this discussion. You might then consider ingress studies, solar eclipses (lunar for London and Asia equity indexes). Then look at the specific aspects: opposition and conjunction between Jupiter and Pluto or Saturn and Pluto on the Dow Jones Industrial Average from 1900 offers an interesting place to begin. That will be enough to get you started. It will also leave you eager to become a more serious student of astronomical events and cycles as utilized by W.D. Gann.

In **Figure 6.8**, the astronomical angles using the Fibonacci numbers 5 and 34 remain on the 2-month EUR/USD chart. To create a more meaningful confluence target on the time axis, a trine or 120-degree angle between the planets Jupiter and Venus has been marked with a vertical line. The angle of separation between planets is called an *aspect*. The aspect cycle does not occur in a fixed interval period. Traders in our seminars often want to jump right to the study of time confluence zones. Be careful, you need command of price confluence targets first, and only then begin to study confluence analysis of time cycles. We think we can control events

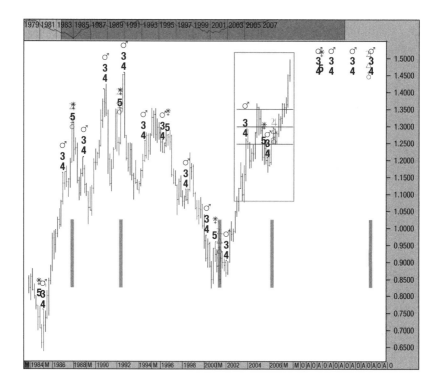

FIGURE 6.8   EUR/USD 2-Month Chart

Connie Brown, www.aeroinvest.com. Source: Copyright © 2008 Market Analyst Software

and their timing, but this type of work shows we don't control very much at all.

Market Character between Confluence Zones

Throughout this book you have seen several methods to develop confluence targets in both price and time. Now it is important to clearly grasp how the spread between confluence zones along the x-axis or y-axis influences how markets move.

We will look more closely at market action between horizontal price zones next. Ford Motor Company (F) is displayed in a weekly chart in **Figure 6.9**. This chart will help us look more carefully at the

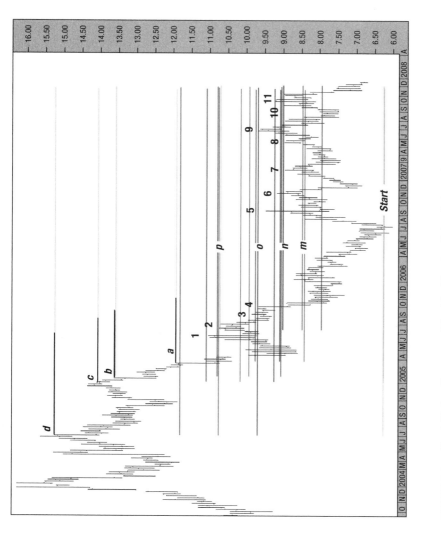

FIGURE 6.9 Ford Motor Company Weekly Chart—Price Internals

Connie Brown, www.aeroinvest.com. Source: Copyright © 2008 Market Analyst Software

price internals within a chart. This is something very few people do until they have traded a long time.

Remember, if we want to create levels of resistance over a current market price, the start of all ranges will always be from a price low, and then sequential bar highs will end the ranges. Near the price low bottom a level marked *start* shows where all the selected ranges that were subdivided begin. The actual price low is not used as a key reversal is present, and the market close and open does not fall under the defined start line. There are four ranges in this chart and all four end at strong bars marked at levels *a*, *b*, *c*, and *d*. None of the endings of the ranges are at major swing highs. This will be the case most of the time.

I know these ranges are correct for several reasons within the market data. When the ranges are subdivided into the ratios 38.2 percent, 50.0 percent, and 61.8 percent, four confluence zones develop in tight bands marked *m*, *n*, *o*, and *p*. One of the confluence zones, *n*, is reinforced with three Fibonacci ratios within this zone. Confluence zones with multiple ratios, developed from gaps, or extremely strong bars all have a higher weighting than two ratios from lesser points. In this chart, confluence zones *n* and *p* are of greatest importance.

Starting from the pivots marked *1* and *2*, the move into pivot *1* manages to push through confluence zone *p*, but fails at minor resistance under pivot *1*. Then when the market falls back and makes a new challenge it fails at pivot *2*, right at confluence zone *p*. We know these ranges are being respected by this market. In other words, we have found the price grid this market is using to develop future swings.

The market falls and then uses pivots *3* and *4* in a very similar manner as pivots *1* and *2*. But the market is much weaker, and a third wave decline begins from pivot *4*. The market makes a hard fast break through levels *n* and *m*. These confluence zones are all levels of resistance. We defined the ranges from the same start at the bottom of the data. *Do not use them as support.* If we wanted to know support, we would have started all the ranges from a high.

If you sold short into confluence zone *m*, stops would have to be at least over zone *n*. You cannot place stops between zones. (Assume you will be filled at the next zone.) To create a risk-to-reward ratio, you would use the range *dp* that ends at level *d* with the midpoint confluence zone at *p* to create an equality swing. The measured range is then projected down from confluence zone *p*. You would find the target was precise at the chart lows. If you sold level *m*, the risk-to-reward ratio of 1 : 3 would be present using the price projection technique just described. If that was the criteria to establish a trade, it is important not to change the strategy. Your profit would have been banked before the start of the market move towards pivot *5*.

From the market low, a strong rebound develops directly to the first confluence zone at *m*. This is common market action. Then a pullback follows that resumes the move up towards pivot *5*. While pivot *5* blew through the confluence zone at *n* and through a minor level of resistance just above it, the price data failed to reach confluence zone *o*. This is very significant because the key reversal at pivot *5* had no interference or reason to stop before confluence zone *o*. It is a sign the brief uptrend is exhausted. The market pulls back and then moves up into confluence zone *n* at pivot *6*. While confluence zone *n* is exceeded, the key reversal fails at the minor level of resistance and then immediately falls back to confluence zone *m*. This is why making these minor levels of resistance or support a soft gray or faded color in the background can be of value. It is also important to notice bar opens or closes do not occur over confluence zone *n* near pivot *6*.

Pivot *7* fails under confluence zone *n*. The market pulls back and fails at pivot *8*. This is a weekly chart. Your oscillators will warn you the market has the strength to create the swing up into pivot *9* in the weekly time horizon, but the monthly chart will tell you the swing up is corrective. Always use two different time periods to make an opinion on how a market will respond to a confluence zone. In this weekly chart, Ford fails at confluence zone *9* without exceeding it. A decline then follows, and you should be able to read pivots *10* and *11*, including the small matching shoulder to the right side of *11* under zone *n*. Notice how the price data moved between zones *m* and *n* from pivots *7* to *10*. All of this price data action confirms the market is working within the grid created within this chart.

In the next example in **Figure 6.10**, we will examine a 1-week chart for CBOT Wheat Futures to mix up the market examples. Wheat offers an ideal example of how markets that form very strong rallies will break hard when they correct. These markets will create wide

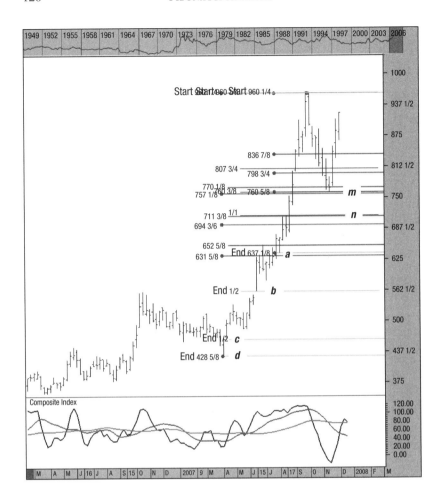

FIGURE 6.10 CBOT Wheat Futures 1-Week Chart

Connie Brown, www.aeroinvest.com. Source: Copyright © 2008 Market Analyst Software

empty spreads between the developing confluence zones. The target zone was level *m*. It is possible the confluence zone at level *m* is only the midpoint of this rally. You can use the box method described and measure the range from *d* to *m*. Then project the same range up from *m*. That is a big move, so you would further subdivide the projected box range from *m* to a new high into the Fibonacci ratios.

The high was used to start the ranges that end at levels *a*, *b*, *c*, and *d*. The reason the high was used is the most recent price bar is higher than the bar behind the key reversal market top. So all levels of resistance have been broken, and only the market top remains. The range ending at *b* is an important gap discussed early in this book. Always use gaps in this manner to define the ending of one of the selected ranges. You should also notice that the close of the bar after *b* is near the high two bars behind *b* that precedes the key reversal just behind *b*. Confluence zone *m* held the market decline from the high, but the spread and empty void between *m* and *n* means this was not the place to go 100 percent long. You might consider a 30 percent long position for a long horizon trader. When the market tests 770⅛, add 50 percent. This is the second bar past the low at *m* that tests a single support line viewed as minor support. This would be a safer entry than at the pivot low on *m*. Then, as the market runs past the high of the bar that tested the zone at *m*, the final 20 percent might be scaled into the long horizon position. Each leg should be treated as its own position and not averaged. Therefore, each traunch has its own risk-to-reward ratio and stop loss trigger. As the confluence zone at *m* can be viewed as the midpoint from which to make future Fibonacci expansion targets, it can be said this market will definitely exceed the current high. Know where you would be wrong to hold such a bullish opinion. If the market breaks the confluence zone at *m*, before reaching the equality swing projected, a new market study must be made. You would also make this evaluation while you were out of the market and flat.

The next market example has been chosen deliberately because of its price complexity. **Figure 6.11** is a 5-week bar chart of the 30-Year U.S. Treasury Bond futures. This market is choppy in nature as it is defining a rising wedge that develops many back-and-fill swings.

The start is important for defining the ranges in this chart. The swing into the actual market high has nearly been fully retraced. *Take the most current swing high because it has broken all resistance levels behind it by nearly defining a full retracement and double top.* It is not a double top, but it tried hard to create one before falling to pivot *1*. Truncate the spike and set the starting level where numerous bars have failed. There are six ranges and the lows that define the endings have been labeled *a*

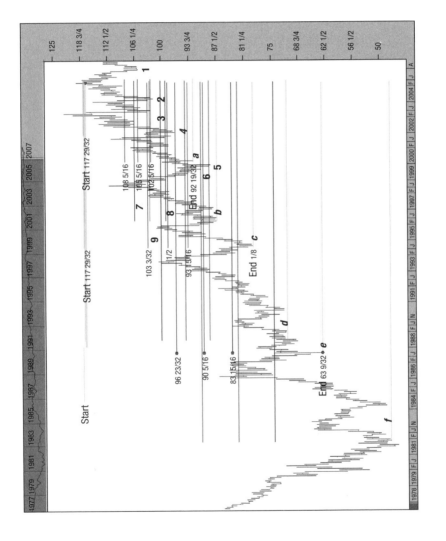

FIGURE 6.11 30-Year U.S. Treasury Bond Futures—5-Week Chart

Connie Brown, www.aeroinvest.com. Source: Copyright © 2008 Market Analyst Software

128

through *f*. This time I have not labeled the zones so you can examine the placement on your own. But use the zone nearest pivots *2* and *3* to set your eye for what you should define as a confluence zone. It does not matter if it is a 1-minute chart or a semiannual chart; the look and feel of a confluence zone within any chart will be similar. As you study pivots *1* through *9*, consider the market's reaction to these past swings. If the market respected these levels in the past, you have set the correct grid for the future. The horizontal lines were not extended to the right. This is a working chart for this market, and where they end on the right was the current bar when they were created. Then pivot *1* fell into a support area that would be clearer if the Fibonacci zones were extended. This chart cannot be used to create resistance unless you develop Fibonacci expansion targets or drop down to a shorter time horizon to use pivot *1* as the start of a range. It would be better to create expansion targets. (We will review this technique in the gold market in the next example.)

The zone near pivots *2*, *3*, and *7* is clearly important for any expansion work developed. The next zone would be just above the label *a*. If you can duplicate these zones on your own in a monthly chart, you are doing extremely well. The key to selecting the ranges in such a chart is to end the first range at the highest price low first. Then scan downwards through the data to sequential lows. Do not jump down to the extreme low because you will not be able to see the internals for all the other ranges you need to consider. By now you have seen I rarely use the price low. When the ranges are subdivided, you may not want to see the secondary minor levels. The solution is to use a horizontal line to mark the widths of the confluence zones and then remove all the Fibonacci subdivisions. The upper and lower boundaries of confluence zones are essential, as you will need the zone width for trading strategy. When you can delete the other calculations leaving just the confluence zones, they can be used for additional analysis techniques. A few of these applications follow in the last two charts.

Figure 6.12 is a 2-month chart of COMEX Gold futures. The three different Fibonacci expansion projections made in this chart define a confluence target at *n*, *o*, and *p*. The current market highs are currently just short of confluence zone *n*. The price data consolidated for several months under the confluence zone that formed at *m*. The

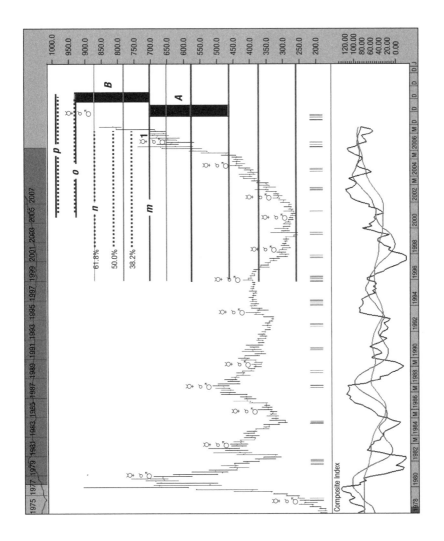

FIGURE 6.12 **COMEX Gold Futures**

Connie Brown, www.aeroinvest.com. Source: Copyright © 2008 Market Analyst Software

130

respect shown to this resistance area is marked *1*, which means this zone will be an important marker for future market swings. Take a look at the oscillator. The Composite Index[2] has diverged from price highs, but we are justified in asking if target zone *n* or *o* is the more probable target. Adding to our concern is seeing the oscillator pressing under its own two moving averages as the shorter-period average crosses down through the slower moving average. The market is clearly overbought, but be very careful about zone *n*. Time analysis shows zone *p* will be realized before a larger correction develops.

Both target zones *n* and *o* in Figure 6.12 are confluence zones containing Fibonacci expansion projections. But zone *n* is not as strong a confluence zone compared to *o*. The zone at level *o* has much greater weighting because it includes an equality swing target of range *B* derived from the range marked *A*. Range *A* begins where the strongest start of the rally ignites upwards and ends after the market showed months of respect to the target confluence zone at *m*. Please do not ever forget that the most significant points, or milestones, the market will give us are gaps and points that mark the start of strong forceful moves. This is true if you are working with intraday or longer-horizon charts. In this case, the start of range *A* is the start of a third-of-third rally, using Elliott terminology. In other words, it is always the point at which the market has everyone on the same side. The shorts have to get out, the early longs want to add to their positions, and the laggards are desperate to jump in. In declines, the same sentiment occurs at these market junctures. Only the Elliott Wave Principle gives us the language to map and discuss milestones of market sentiment, making this methodology a valuable addition. Level *n* should be viewed as minor resistance, and level *o* as major resistance where a corrective pullback or pause could occur before the trend resumes upwards.

In the last two chapters you will learn why various Fibonacci confluence zones like levels *n*, *o*, and *p* are not of equal weighting or importance. Before long you will ask, "Which zone is more significant to the one that began the entire price move?" That is when you will know you are ready to do more than just create the confluence zones. When you start asking which zone is the most important, you are beginning to study how these confluence zones relate to one another. In Chapters 7 and 8 we will look at how proportion is evaluated so we can solve problems involving ratios. Fibonacci analysis is really just a subset within a larger field of analysis that is new to the industry involving harmonic intervals. But do not move to this level of study until you are proficient in working with confluence zones by themselves.

The Hong Kong WEB on the NYSE (EWH) is displayed in a 2-week chart in **Figure 6.13**. Figure B.2 in Appendix B is the same chart before the Gann fan angles where added, if you want to examine how the Fibonacci confluence zones were created without the clutter. Gann angles radiate forward in an upwards or declining direction with a specific slope. When a line is created with a growth rate from a price pivot point with one unit of price relative to one unit of time, it creates a 1-by-1 trend line at a 45-degree angle. The precise angles that define a complete Gann Fan are 3.75°, 7.5°, 15°, 26.25°, 45°, 63.75°, 75°, 82.5°, and 86.25°. Figure 6.13 displays a few Gann angles to demonstrate that many technical tools perform with greater accuracy when you use the same pivots that create range extremes for confluence zones. In Figure 6.13 the intersection of two Gann angles from different confluence zones define a major market pivot high. This is not a final top, but accurately warns that this market is at an important level of resistance. Notice how the market fell to the first confluence zone near 19.90.

If you are new to these methods, your hands are full learning the basics, and you need to practice in many different market charts. Use old data at first so you can scan forward to see how accurate you are becoming. Listed in Appendix B are a few common errors traders I've taught encounter; the aim of that appendix is to make you aware of these common difficulties. Although you might find it difficult to look at the smaller details within a chart, you need to recognize that developing an ability to do this is absolutely essential. The traders I've worked with who struggle at first generally make broad-stroke assumptions far too quickly and have never allowed themselves to look at the rich source of directional information forming within the internal price structure. The internal structure will not get in your way of making a decision; it will increase your probability of being right.

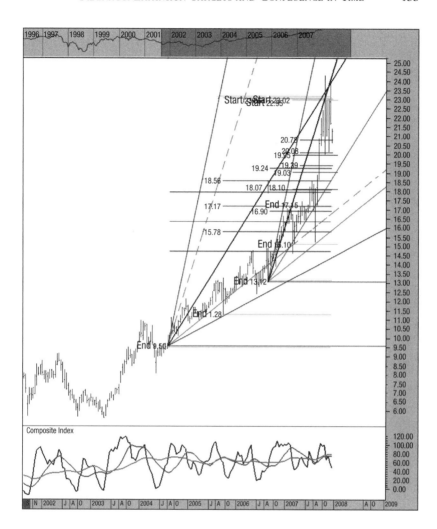

FIGURE 6.13   Hong Kong WEB (EWH) 2-Week Chart

Connie Brown, www.aeroinvest.com. Source: Copyright © 2008 Market Analyst Software

There will come a point in your trading when you are amazed at how the market respects confluence zones within any time horizon. These methods will become fluid and effortless with practice. It will not matter what market you are asked to consider. I once found an exact pivot deemed impossible for the Thai bhat/lira currency cross. I just used these methods and switched the chicken track data to a line chart to make the calculations. The methodology and logic tree does not change and is a sound approach for any market you may wish to trade. The caution is to never study one time horizon or one market alone. Always find the correlated, or inversely correlated, market to help guide you. If one market blows through a zone, it is an early warning about another.

We are ready to dig deeper to explore how these confluence zones relate to one another. Some of you may leave at this point, and I wish you trading success and the patience to learn these methods. They have defined a career that built my home and supported a dream. I am grateful to Joe DiNapoli who first introduced the concept of Fibonacci confluence to me many years ago. But then I took the idea of confluence to an entirely different level and with experience new methods began to emerge. For over ten years people in the industry have asked me to write this particular book. I guess I never felt ready before now because it takes years of experience to finally feel you have the confidence and security to reveal what others do not discuss. There is no "bs" within these pages, just an open record of what has been proven to be a working method that I know has held up over years in the most volatile of markets and over changing times. Let other methods give you permission to act upon the target zone. Don't read indicators alone. If you are looking for more behind these methods, the final chapters will give you insight into how the confluence zones relate to one another. You will discover that the proportions between zones are harmonic. This opens a different kind of chart analysis, and it is interesting how life comes full circle. The discovery of harmonic relationships between confluence zones brings us back to the Pythagorean doctrine that all things are connected by harmonic proportions. I believe this to be true now. For the few brave enough to read on, we are ready for those final discussions. For those who need more time to digest the methods described, my best wishes to you for trading success, and may success bring you peace.

CHAPTER NOTES

1. Using a 34-degree progression is very similar to what is known as an *ingress cycle*, where a planet crosses a house cusp. Ingress studies are recommended, but a Fibonacci number has been deliberately selected so those of you more familiar with astrology will broaden your research to include astronomical progressions using the Fibonacci number series for angle analysis in addition to your other work.

2. The Composite Index Oscillator formula I released in *Breakthroughs in Technical Analysis* (Bloomberg Press, 2007). You will find more about this oscillator and how it is used to warn when RSI is failing to detect a market reversal in my book *Technical Analysis for the Trading Professional* (McGraw-Hill, 1999).

CHAPTER 7

Rhythmic Wave Diagrams

I INTRODUCED RHYTHMIC WAVE DIAGRAMS in Chapter 4, in the discussion of the Roman Colosseum and how Fibonacci ratios form in the horizontal, vertical, and diagonal axes. The flowing arcs visually connect points that have proportional relationships. But a man-made structure is only a copy of the geometry found in nature. Now we are going to see how biology, physics, and proportional geometry come together in ways you might not have contemplated before. This discussion assumes you have made the connection that mathematical ratios and the resulting proportions found in nature are mirrored and replicated within financial market data.

In prior chapters, we discussed how to reveal the confluence zones that a market uses to build future price swings. The confluence zones formed from the cluster of ratios developed from subdivisions of various price ranges. It was important to use certain internal milestones that consistently created confluence zones, such as gaps and the start of strong third waves. The result was a mathematical grid that overlay the entire data set that could be used in various ways to forecast future measured moves. Now, we are about to find that nature creates a grid of confluence zones from internal structures as well. The conceptual conclusion you need to derive from this chapter so you can grasp the final discussion is this: The points where multiple ratios intersect are not related to one another in a simple linear series. Because markets mirror the laws of nature, you can deduce that the relationships

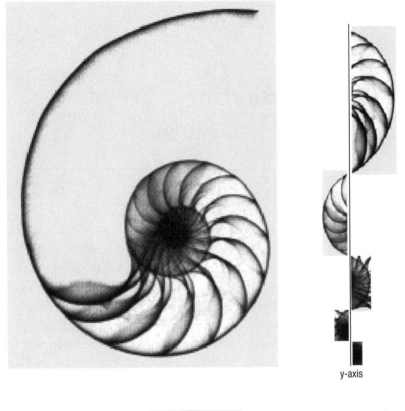

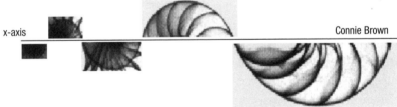

FIGURE 7.1   Nautilus Shell Geometry

between Fibonacci confluence zones within market price data do not create linear series either.

The last diagram in Chapter 4 revealed the geometric figures that evolve from within a nautilus shell. **Figure 7.1** demonstrates the golden spiral form of the nautilus shell as well. While the geometry was revealed, the question of how to map a spiral and the numerous geometric figures onto a two-dimensional chart was left unanswered. Now is the time to address this question.

Traders familiar with Fibonacci ratios within nature are well aware that plants, fish, and butterflies all exhibit 161.8 percent and 61.8 percent ratios within their structures. It is hard to write a book about Fibonacci ratios without some sort of bug or plant as a point of discussion. But this time our goal is a deeper discussion, rather than mere proof these ratios exist in nature. Now is the time to consider how the points that form the extreme and mean proportion relate to one another.

Notice that I have mapped the skeleton of a frog using a rhythmic wave diagram in **Figure 7.2**. The last time I looked at a frog this closely was in high school biology, and I hate to admit how long ago that was. But this rhythmic wave diagram will help you see how examining the frog's proportional sections will further our discussion about the relationships between confluence price zones.

The rhythmic wave diagram on the right side of an actual frog's skeletal form details the proportional connections along line *sm*. View these connection points as confluence zones. Line *sm* is the maximum stretch between the longest finger above the forearm and the longest digit of the foot. The rhythmic waves or confluence zones connect the primary joints and body extremities. As we study the internal string of subdivisions along the line *sm*, it becomes clear a logarithmic proportional relationship is present that connects the smaller parts within the whole. The analysis of greatest interest to us is how the Fibonacci ratio 0.618 appears within this proportional diagram.

Observe the wave proportion between confluence points *s*, *n*, and *m*. You will find a sine wave that begins at point *s* and swings out to the left. It then cuts back and swings through line *sm* at point *n*, which aligns with the frog's knee. The proportional wave continues and swings out to the right and reconnects at point *m*. For each

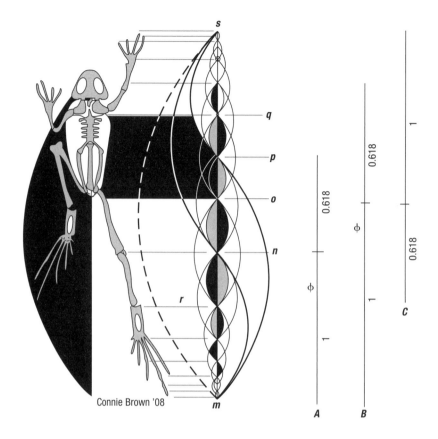

FIGURE 7.2   **Frog Skeleton with Rhythmic Wave Analysis**

Source: Connie Brown, © 2005 Aerodynamic Investments Inc., www.aeroinvest.com

---

proportional wave there is a mathematical relationship between the connected ranges, *but not all segments exhibit Fibonacci ratios.* This last point is extremely important. It can be suggested that Fibonacci ratios are only part of a larger subset.

If we want to find the anatomical joints or extremities that form a Fibonacci ratio, we have to break away from measuring the whole range *sm*. For example, the confluence zone at point *p* aligns with the frog's abdominal joints. If we measure the distance from *p* to *m* with a proportional ratio compass, we will find that the Fibonacci ratio 0.618 falls right at the knee, at point *n*. Proportional line *A*, on the far right of the diagram, will help you visualize this extreme and mean proportion. Study the joints at *p* and *n* very closely, as if you were an engineer evaluating thrust points of stress. The point that would bear the greatest force as a frog leaped forward would be the joint at the knee, which would create the force that is jammed into the abdominal joint at *p*. The joint at *p* creates an angle at which the bones must transfer the thrust to the upper portion of the backbone and upper body frame.

Had we measured a range from the head to the extreme at *m*, the proportional Fibonacci ratio would fall at point *o*, which aligns with the frog's tail. Anatomical physics makes more sense to connect the Fibonacci ratio with the head and body extremes than it would to connect the head with the abdominal joints. The more you study these ranges and consider mechanical physics in light of the anatomy and its function, the more astounding the diagram becomes.

If you measured the range *sm* as 1.618, you would find that no confluence zone or joint falls on the 38.2 percent or 61.8 percent ratio. Only when you measure a range that makes sense in terms of anatomical mechanics or body proportion do you find the intersecting body joints with Fibonacci ratios. Fascinating. Proportional lines *A*, *B*, and *C* on the far right of the diagram map only three extreme and mean proportions. There are several others. For example, use a proportional divider to measure the length of the body from the top of the head to the tail. Then find the 0.618 relationships within the body.

Consider the proportional math found in nature further. Point *o*, which defines the tail end of the frog's body, has no proportional connection to points *p* or *n*. However, point *o* is a link between these two anatomical points that logarithmically connects the whole.

Measure from the ankle and point *r* to *s*. The only joint creating the extreme and mean proportion resides at point *o*. Now consider the distance of line *ro* and line *np* and that mechanical thrust is dissipated between each set of points. While my knowledge of body mechanics is only related to my experiences as an international swimmer years ago when computer analysis was becoming common, I understand enough to recognize that proportional math shows that each logarithmic section is important to the whole, but some proportions have greater weighting due to specific mechanical functions. The key point is that mathematical relationships within the skeleton of the frog are not simply linear relationships; nature is not linear. The ratios of greatest interest do not fall one after another in a linear sequence. As markets follow laws found in nature, *the relationships between the confluence zones are not linear either.*

What relationship, if any, exists between the confluence zones that do not create extreme and mean proportion in a logarithmic series? *It is important to realize proportion is used to solve problems involving ratios.* We must push ourselves beyond the mere identification of confluence zones, and venture into a new field of proportional analysis that shows that harmonic intervals develop within the grid of confluence zones. We will look at this new field of study for chart analysis in the final chapter.

CHAPTER 8

Harmonic Unity Within Market Price and Time

THROUGHOUT THIS BOOK we have subdivided price ranges and discussed how markets respect the zones in which these Fibonacci ratios cluster together. In this chapter you will study how these confluence zones relate to one another. To do this, you need to have a deeper understanding of how proportion is evaluated so you can solve problems involving ratios.

You will recall from an earlier chapter that proportion is a repeating ratio that typically involves four terms, 4 : 8 :: 5 : 10, stated as "4 is to 8 as 5 is to 10." The Pythagoreans called this a four-termed discontinuous proportion. The invariant ratio here is 1 : 2, repeated in both 4 : 8 and 5 : 10. An inverted ratio will reverse the terms, so it can be said that 8 : 4 is the inverse of 4 : 8, and the invariant ratio is 2 : 1.

The Pythagoreans knew of three different ways to identify proportional relationships between elements. These remain the three most important ways for us to study proportion: arithmetic, geometric, and harmonic.

Arithmetic Proportion

A 2-week chart of Centex is displayed in **Figure 8.1**, as it is easier to visualize data ranges than it is to study algebraic equations. Notice that the ranges drawn in this chart begin and end with values, as we have discussed, that create confluence zones. Do not try to read a market projection method into this discussion.

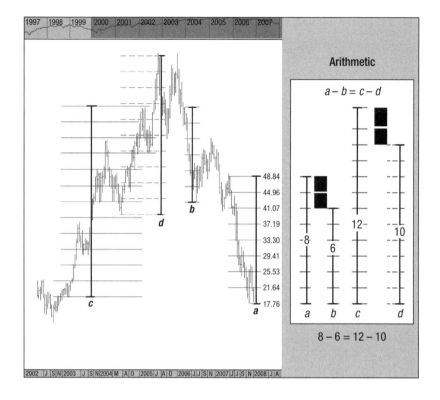

FIGURE 8.1   Centex Corp. (CTX) 2-Week Chart—Arithmetic Proportion

Connie Brown, www.aeroinvest.com. Source: Copyright © 2008 Market Analyst Software

You will see that range *a* is subdivided into 8 equal units. The length of 1 unit within range *a* allows you to see that price range *b* is 6 units in length, range *c* is 12 units in length, and *d* is 10 units in length. The algebraic expression to define arithmetic proportion is a − b = c − d. *The difference between the first two elements must be equal to the difference of the second pair of elements.* The ratios are not the same; that is to say, *a/b* does not equal *c/d*. But the arithmetic differences are both 2 units as illustrated with black boxes on the right-hand side of Figure 8.1. An arithmetic proportion would not have been demonstrated if range *c* had been measured from the price pivot low behind it. Similarly, you would not have an arithmetic proportion if range *b* had ended at the price high. Using internal structures, as we have done throughout this book, reveals an arithmetic proportion that would have been disguised with a surplus or shortage if we had assumed the price highs and lows were always the most significant pivots to use. If you think about it, a simple series of whole numbers, such as 1, 2, 3, 4, 5, is an example of a continuous arithmetic proportion because the difference between each whole number is always 1.

Geometric Proportion

While the arithmetic proportion only has equality in the differential comparison of the elements, a geometric proportion must have a proportional ratio in the range segments and create a ratio between the differences. The algebraic equation for geometric proportion is a : b = c : d. This is the same as using a fraction that is written as $a/b = c/d$. The nominators and denominators can be divided or multiplied to easily find geometric proportion. For example, you can create the ratios of 4 : 8 and 2 : 4 from **Figure 8.2**, viewing the left-hand side of the diagram. The ratios of 4 : 8 and 2 : 4 are divisible, and they can be written 4 is to 8 as 2 is to 4 (4 : 8 :: 2 : 4). Geometric proportion allows the elements to be checked by multiplying the inner elements and the outer elements for equality comparison. For example, $8 \times 2 = 4 \times 4$.

You may recall from the first chapter that the geometric mean between two numbers is equal to the square root of their product. Therefore, it can be said that the geometric mean of 1 and 9 is $\sqrt{(1 \times 9)}$, which equals 3. This geometric mean relationship is written as 1 : 3 : 9, or inverted, 9 : 3 : 1. The 3 is the geometric mean held in common by both ratios. The Pythagoreans called this a three-termed continuous geometric proportion. The proportion of the Golden Ratio is a three-termed continuous geometric proportion.

Harmonic Proportion

Arithmetic proportion is about the differences between elements. Harmonic proportion is about the ratios of the differences. In Figure 8.2 the harmonic proportion is illustrated on the right-hand side of the illustration, which shows the resulting harmonic ratio of 1 : 2. If four

Geometric

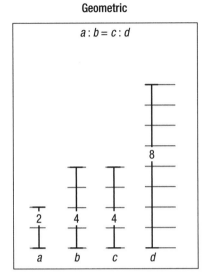

Harmonic

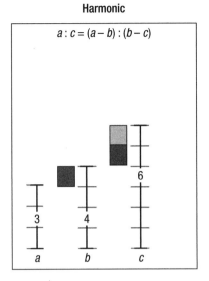

FIGURE 8.2   Geometric and Harmonic Proportions

Source: Connie Brown

elements had been shown, the middle elements would always be the same number. Harmonic proportion is the mathematical ratio of musical intervals. Developing a better understanding of harmonic proportion and the mean relationships are most pertinent to building upon our prior discussions in this book, because harmonic relationships connect the major confluence zones in our market data.

Harmonic Intervals

The mathematical comparison of market price swing lengths is conceptually no different from the work of ancient Greek and Arabian geometricians who studied string length ratios to define musical intervals. The Greeks primarily used verbal terms to describe the mathematical ratios of musical intervals. Furthermore, the Greeks never employed fractions such as 1/2, 2/3, 3/4, and 8/9, to identify musical intervals. The ratios of Greek music always express quotients that are

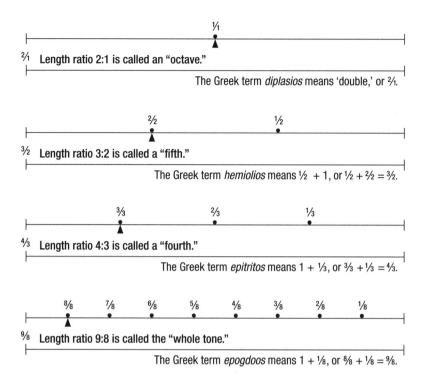

FIGURE 8.3   Musical Intervals Expressed as String Length Ratios

Source: Connie Brown

greater than one. For example, in **Figure 8.3** the ratio of the octave (or *diapason*) is *diplasios*. Since the prefix *di* means "two," and *plasma* means "something formed," *diplasios* means "twofold," or "double." Therefore, the octave is a musical interval produced by a string that is twice as long as a similar string; hence, the resulting length ratio is 2/1 or 2 : 1. Similarly, the term for the ratio of the fifth (or *diapente*) is *hemiolios*. Since the prefix *hemi* means "half," and *olos* means "whole," *hemiolios* means "half and whole": $1/2 + 1 = 1/2 + 2/2 = 3/2$. The Greek term for the ratio of a fourth is *epitritos*. Since, in a mathematical context, the prefix *epi* describes the operation of addition, and *tritos* means "third," the term *epitritos* denotes "one third in addition," and connotes "one and one third": $1 + 1/3 = 3/3 + 1/3 = 4/3$. Therefore, the fourth is a musical interval produced by a string that is by a third longer than another string. Similarly, the term for the ratio of the whole tone is *epogdoos* (or *epiogdoos*), which connotes "one and one eighth": $1 + 1/8 = 8/8 + 1/8 = 9/8$. Therefore, the whole tone is a musical interval produced by a string that is by an eighth longer than another string.[1] It was Nicomachus[2] who abandons the concept of length ratios to study harmonic intervals in Chapter 6 of his *Manual of Harmonics*.[3] He tells the story of Pythagoras comparing various weights suspended from the ends of strings and discovers that the harmonic intervals for the octave, fifth, and fourth vary directly with the tension of the string. The octave was produced from the sound of two similar strings of equal length when one was pulled by a 12-pound weight and one by a 6-pound weight. The tension on the strings created the harmonic interval. The ratio is 2 : 1. The fifth was duplicated by weights of 12 and 8 pounds, or 3 : 2. The fourth was created from 12- and 9-pound weights, or 4 : 3. You will recall it was Pythagoras's study of hammer weights striking an anvil that led to the discovery of the ratio 1.618. As market technicians very familiar with the ratios phi and Phi, we are not venturing that far from the first apple tree where these concepts originated.

Greek Harmonic Intervals, Tetrachords, and Scales

Musicians who played Greek lyres tuned the four strings to tetrachords,[4] or to simple scales emanating from the interval 3 : 2, or the fifth. Ptolemy (*c.* 100–165 AD), and Archytas (*fl. c.* 400–350 BC) identified three different kinds of tetrachords: the *diatonic genus*, the *chromatic genus*, and the *enharmonic genus*.[5] These tetrachords can be found in a treatise entitled *The Elements of Harmony* by Aristoxenus[6] (*fl.* fourth century BC).

The intervals of greatest value to the ancient Greeks, such as Philolaus, are the intervals of value to us today in analyzing market swings and Fibonacci clusters. Philolaus defines[7] three intervals:

Syllaba = 4/3, *Di'oxeian* = 3/2, *Dia Pason* = 2/1

He went on to describe the following mathematical harmonic intervals or relationships that help us see matrix math being applied to further the capability of working with harmonic intervals:

$$\text{Harmonia} = 4/3 \times 3/2 = 2/1$$
$$\text{Epogdoic ratio} = 3/2 \text{ divided by } 4/3 = 9/8$$
$$\text{Two whole tones} = 9/8 \times 9/8 = 81/64$$

Philolaus called the smallest interval of the diatonic genus, ratio 256/243, a *diesis*. With the exception of Philolaus, all ancient Greek and modern theorists refer to the ratio 256/243 as a *limma*, or the "remainder." This is the interval that remains after one subtracts two whole tones from a fourth.

You may have realized that the harmonic intervals between two notes (or confluence zones) form ratios that utilize the first seven numbers of the Fibonacci number series (0, 1, 1, 2, 3, 5, 8). Examples of the musical notes with fixed harmonic relationships to one another are listed in **Table 8.1**.

TABLE 8.1   Harmonic Intervals from Fibonacci Numbers

Fibonacci Ratio	Perfect Frequency	Note in Scale	Musical Relationship
1 : 1	440	A4	Root
2 : 1	880	A5	Octave
2 : 3	293.65	D4	Fourth
2 : 5	174.6	F3	Aug. Fifth
3 : 2	659.4	E5	Perfect Fifth
3 : 5	261.7	C4	Minor Third
3 : 8	164.8	E3	Fifth
5 : 3	739.8	F#5	Sixth
5 : 8	277.2	C#4	Third
8 : 3	1174.60	D6	Fourth

Harmonic Series

We are interested in more than the interval between two notes. That would be like having one major price swing in place off a bottom and the first pivot, then not knowing what will happen next. But market swings form a series of harmonic intervals. This is how a harmonic series is developed: A fundamental frequency is a selected vibration, such as a tuning fork that vibrates or oscillates at precisely 261.626 Hz per second. This frequency in Western music is known as the note middle C on a piano, or C4 in scientific pitch notation. The piano has a range of seven octaves plus a few extra notes, and the human ear can hear nearly eleven octaves. The frequency of a vibrating string at 523.251 Hz is one full octave above middle C (C5), and it is vibrating at twice the rate of the lower C (C4), so the harmonic interval of an octave is 2 : 1. Picture a fundamental frequency in your mind as a price milestone like the start of a strong move, a gap, or a significant market pivot. Conceptually, you might also consider 261 Hz to be the price low of the 1987 crash in the S&P 500. Now we want to know what Fibonacci confluence zones have a harmonic proportional relationship to this market pivot so we can identify mathematically where the next major market reversal may occur if the market is building upon the same price grid.

Frequency is represented with the letter f, so f equals 261 Hz in my example. Music is a logical matrix-based notation. If $f = 261$ Hz, then the harmonic series created from this fundamental frequency of 261 Hz is $f^2 = 522$, $f^3 = 783$, and so on to f^n. A harmonic interval is the ratio of the frequencies of two tones. For example:

$$f^3/f^2 = 783 \text{ Hz}/522 \text{ Hz}.$$

The higher of the two values is placed over the lower because we are considering an upward harmonic interval. To complete this example,

$$
\begin{aligned}
f^3/f^2 &= 783 \text{ Hz}/522 \text{ Hz}, \\
&= (3 \times 261 \text{ Hz})/(2 \times 261 \text{ Hz}), \\
&= 3/2 \text{ or, written as the ratio, } 3 : 2.
\end{aligned}
$$

We have already visually experienced a perfect fifth, or the ratio 3 : 2, in the geometry of the nautilus shell. We have also experienced a perfect fourth, which has the exact ratio of 4 : 3. But now that you have an appreciation of the significance of these ratios, they are best illustrated in a more practical way as they occur within a monthly chart for the S&P 500 in **Figure 8.4**.

The S&P 500 chart in Figure 8.4 shows that the nearby highs at the end of 2007 are pressing up against more than just a zone of resistance. The three most critical pivots precisely connected by harmonic proportion in this market are the 1987 low, the 2002 low, and the nearby highs. The current market high is an exact perfect fifth harmonic interval relative to the 2002 price low. The ratio of the 1987 S&P 500 price

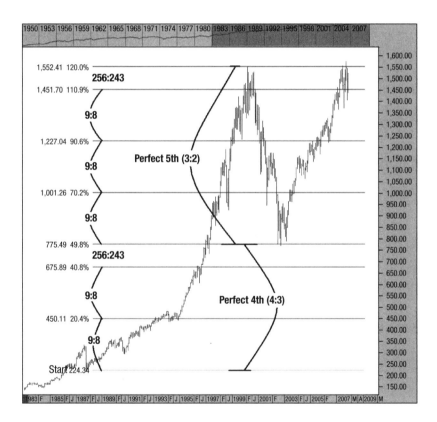

FIGURE 8.4   **S&P 500 Futures Respecting a Harmonic Perfect Fourth and Perfect Fifth Interval**

Connie Brown, www.aeroinvest.com. Source: Copyright © 2007 Market Analyst Software

low and the 2002 price low is an exact perfect fourth harmonic interval. When you want to know the mathematical start of a market move, use the harmonic proportion as the relevant milestone. Therefore, we should note that the S&P 500 market has placed great importance on the 1987 price low and the current market highs, suggesting this is a high-risk pivot zone for a market correction. This harmonic geometry added to our technical indicators becomes a powerful tool for us.

While Figure 8.4 shows the S&P 500 futures respecting a harmonic perfect fourth and fifth interval, the chart is showing much more. The juxtaposition of harmonic intervals throughout the chart as marked on the left side of the chart is a specific harmonic series known as the Pythagorean diatonic scale.

Why do we technical analysts and traders have any interest in the Pythagorean diatonic scale? Not only do we find the perfect fourth and fifth intervals within the monthly S&P 500 market, as shown in Figure 8.4, we see that *this market respects all of the harmonic intervals that define the Pythagorean harmonic series.* This harmonic series, then, shows us which Fibonacci confluence zones have the greatest weighting. While not every market will build its harmonic grid according to the Pythagorean scale, this scale is the most important one you will need as you begin your study of market harmonics within North American equity charts.

While most recognizable harmonic music scales are created from a base prime number raised to a series of exponential whole numbers, Euclid created a harmonic scale of fourteen proportions that only needed Philolaus's five different length ratios of 4/1, 2/1, 3/2, 4/3, and 9/8 to create the complete harmonic scale. Plato mentions Philolaus on three occasions in a dialog entitled *Phaedo*,[8] but does not credit him in *Timaeus* when he reveals the diatonic tetrachords.

To suggest we can apply harmonic intervals to market analysis charting and ignore the conceptual "weight ratios" and resulting "vibration ratios" is not unique. Creative theorist-musicians like Euclid, Ssu-ma Ch'ien[9] (also known as Sima Qian) (163–85 B.C.), Ptolemy (*c.* 100–165 A.D.), Al-Kindi[10,11] (d. *c.* 874), and Al-Farabi[12] (d. *c.* 950), simply ignored weight ratios and vibration ratios in their geometric work. All of these writers based their work on the premise that frequency ratios are a function of length ratios. Therefore, it would be correct to apply their conceptual work to price swing and market ratio analysis.

Harmonics show which Fibonacci confluence zones should be given greater weight or importance. Harmonics also help us understand how confluence zones are related, or not related, to one another within our charts. We can examine the S&P 500 monthly chart in **Figure 8.5** to see how matrix math can further enable us to combine harmonic ratios. We can apply the harmonic intervals work of Philolaus in our chart analysis. As an example, the perfect fourth and perfect fifth combine to make one full octave (3 : 2 × 4 : 3 = 2 : 1). This is of

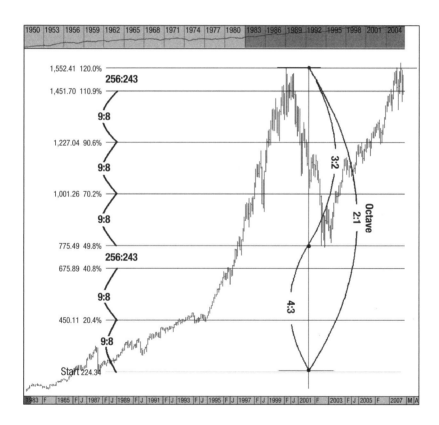

FIGURE 8.5   *S&P 500 Futures—Combining Harmonic Ratios*

Connie Brown, www.aeroinvest.com. Source: Copyright © 2007 Market Analyst Software

tremendous interest to us in current times. The S&P 500 market position at the end of 2007 is at a very critical and significant harmonic inflection point because it is at the full octave interval from the 1987 lows. Only time will tell if the market will respect this zone of resistance in the S&P 500. Oscillators confirm the market is overbought, but oscillators alone cannot tell us the magnitude of the reaction that may follow.

While the Fibonacci confluence zones are not displayed in Figures 8.4 or 8.5, to keep the harmonic scale uncluttered and easy to examine, clearly not all harmonic ratios within a musical scale will mirror the Fibonacci confluence zones we create. We are looking for an overlap between the harmonic series and the Fibonacci confluence zones. In other words, when a harmonic ratio lands within a Fibonacci confluence zone, it is a target area of tremendous interest.

There is more about the Pythagorean diatonic harmonic series to discuss. Notice how, in Figures 8.4 and 8.5, intervals are marked on the far left. The difference between a perfect fifth and perfect fourth $(3 : 2 - 4 : 3)$ is the interval the ancient Greeks called the whole tone, which is the harmonic ratio of $9 : 8$. The Pythagorean scale displayed in Figures 8.4 and 8.5 has five whole-tone intervals and two semitone or halftone intervals. Very quickly a pattern emerges that is best represented with a rhythmic wave diagram. Our earlier discussions about the logarithmic proportions within a frog and about the Roman Colosseum have prepared us to use a more advanced rhythmic wave diagram to evaluate the market.

Figure 8.6 is a complete rhythmic wave diagram, which has become a tool to map all the harmonic intervals within the S&P 500. What you need to conclude from this diagram is that harmonic intervals do not create a linear series. For example, follow the arcs, or oscillations, defining the perfect fifth intervals $(3 : 2)$ within the series. Observe how they must skip over some nodes. The nodes are the arc crossover points through the midline. Notice how a simple rhythmic wave diagram was added to Figure 8.5 to help you see how to read the harmonic intervals in a simple application. Now notice how a few harmonic intervals can only form when the harmonic series begins to repeat beyond the first octave range. The concepts of harmonic analysis apply to both the price and time axes. The idea of fixed-period time

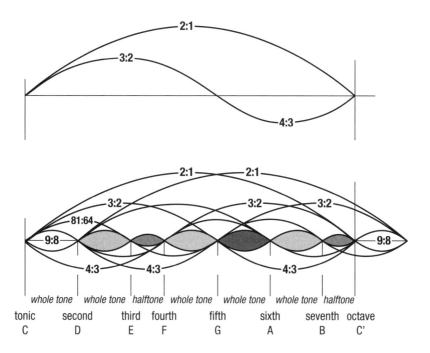

FIGURE 8.6   Rhythmic Wave Diagram

Source: Connie Brown

cycles is shattered by these concepts. Imagine: A cycle low defined as a major price low may not have a comparable cycle low until *x* harmonic units away, which warns you to skip intervals immediately ahead as irrelevant or noise.

The Pythagorean scale and others are created from a prime number in the base that you will find in more detail in a video I donated to the Market Technicians Association.[13] Therefore, further examination of prime numbers is relevant within the context of harmonic analysis.

During a scientific meeting in 1963, the mathematician Stanislaw Ulam began a doodle with similarities to the Square of Nine configuration of a Gann Wheel. The only difference between Ulam's original spiral and a Gann Wheel is Ulam's placement of the number 2 to

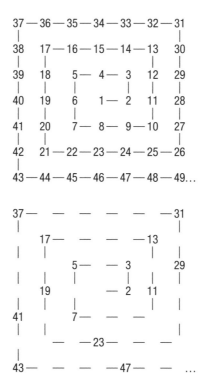

FIGURE 8.7   Ulam Spiral

the right of the first number, as shown in **Figure 8.7**. A Gann Wheel Square of Nine places the number 2 to the left of the first number. As the spiral is created, the number 10 must move one place unit outside the first square, and then the spiral pattern can resume until the number 25, at which point another jump must occur to start the next outer box.

The whole number spiral in Figure 8.7, which mathematicians know as the *Ulam spiral*, was not original, but what Ulam did next was. He began to circle just the prime numbers. On the upper portion of the figure is the number sequence forming the spiral. On the lower portion of the figure, all numbers have been removed *except the prime numbers*.

When we have a computer create this grid from the numbers 1 to 40,000 and then keep only the prime numbers, we begin to see diagonal lines appear in the results, as shown in **Figure 8.8**. This is the hint of another pattern taking form that can only be fully revealed when the computer is then asked to remove all primes, except those that are directly touching two other primes. The remaining primes can touch one another on either side or touch another directly above or below the column or row forming a triangle relationship.

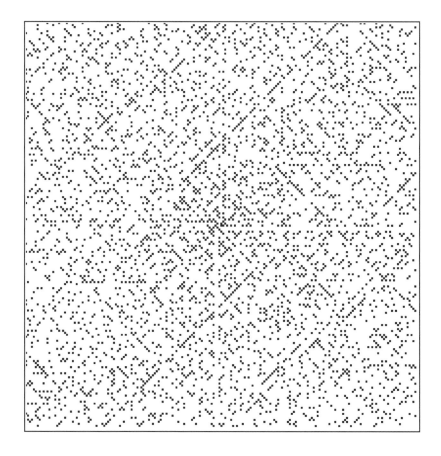

FIGURE 8.8   Ulam Spiral Pattern—Prime Number Pattern

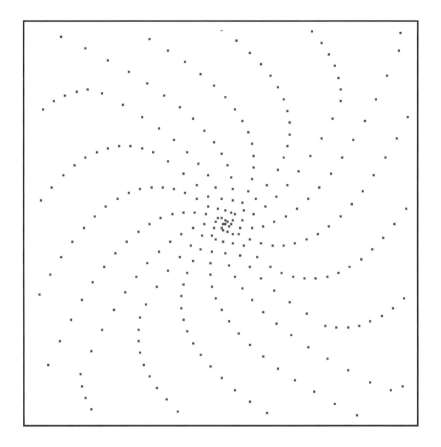

FIGURE 8.9   Ulam Spiral Pattern—Subset of Primes

What remains from the original prime number series is the pattern demonstrated in **Figure 8.9**. As we now know, musical scales have a prime base number raised to an exponential whole number, and we can conclude that Fibonacci numbers are part of a larger mathematical set involving harmonics.

Harmonic intervals form between Fibonacci confluence zones, but not all Fibonacci confluence zones are connected by harmonic intervals or related to one another sequentially. We have only touched upon the study of harmonics within our markets and conceptually opened the door to a fascinating field. The work I am doing will continue,

FIGURE 8.10 **Sunflower Head**

Source: Connie Brown

however, and with your help we will push our industry further into
testing and developing a deeper understanding of how harmonics flow
within markets. Our industry is never stagnant. To do well, we always
have to push ourselves as the markets demand we do. It is my hope
these methods help you come closer to your goals. But I also hope
some of the concepts in these last two chapters begin to drive you a
little crazy so that they may ignite a fire of curiosity and motivate you
to join the research. After all, it is only through a study of harmonics

that we will finally be able to put to rest the burning question, "Why did that market stop exactly on a Fibonacci confluence zone again?!" My best wishes to you.

Connie Brown
December 31, 2007

CHAPTER NOTES

1. Barker, 46–52.

2. Barker, 245–269.

3. Nicomachus, *The Manual of Harmonics* (translated by Levin); Madden.

4. Chalmers.

5. Barker, 303–304.

6. Aristoxenus, *The Harmonics* (translated by Macran), 198.

7. Burkert, 394. "These considerations, along with the archaic terminology, allow us to regard Philolaus's Fragment 6 as one of the oldest pieces of evidence for Greek music."

8. Plato, *Complete Works* (edited by John Cooper), 60d, 61d, and 61e.

9. The oldest source that gives detailed calculations for this twelve-tone scale is a book by Ssu-ma Ch'ien (163–85 BC) entitled *Shih Chi* (*Records of the Historian*), written *c.* 90 BC.

10. Lewis, 225. For a history of Arabian music, read "The Dimension of Sound," by A. Shiloah.

11. Lachmann. Ishaq Yaqub ibn Ishaq al-Kindi (c. 801-873 CE) was the first and last example of an Aristotelian student in the Middle East. He viewed the neo-Pythagorean mathematics as the basis of all science. The oldest source on Arabian music is his work entitled *Risala fi hubr ta'lif al-alhan* (*On the Composition of Melodies*).

❑ Al-Kindi's twelve-tone scale is the first tuning that uses identical note names to identify the tones of the lower and upper "octave." In his text, al-Kindi specifically states that the musical "qualities" of tones separated by an "octave" are identical.

❑ This is the first mathematically verifiable scale that accounts for the Pythagorean comma. In his *ud* tuning, al-Kindi distinguishes between the *apotome* [C sharp], ratio 2187/2048, and the *limma* [D flat], ratio 256/243.

❑ This is the first mathematically verifiable example of a Greek tetrachord on an actual musical instrument.

12. The most famous and complex Arabian treatise on music is a work entitled *Kitab al-musiqi al-kabir* (*Great Book of Music*) by Al-Farabi (d. *c.* 950). See also Farmer (page 28), whose work consists of an annotated bibliography of 353 Arabian texts on music from the eighth through the seventeenth century.

13. Presentation by Connie Brown, Great Market Technicians of the 21st Century: Galileo, Beethoven, Fibonacci. Please contact the MTA at www.mta.org. All proceeds go to the MTA to help rebuild the library.

APPENDIX A

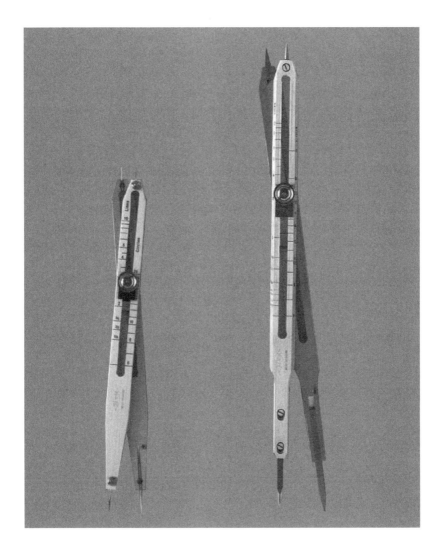

FIGURE A.1

(Left) Alvin 7½ in. Lightweight Duraluminum Proportional Divider, Part Number: ALV-450;

(Right) Alvin 10 in. Proportional Divider, Part Number ALV-458

Manufacturer: Alvin and Company

APPENDIX B

Common Errors

WHILE EVERYONE SEES charts in their own way, there are a few common errors that nearly everyone makes when they attend one of my seminars. This appendix will try to make you sensitive to a few of these common errors so you do not make the same mistakes.

The first example is the chart in **Figure B.1**. This is a 3-day chart for a bank stock on the Australian exchange. The first error is for a trader to start the range to create the Fibonacci retracements in the wrong direction. Start from a high and work down when you want to find support. Work from a low and move up, to determine resistance. In this chart we want to begin to find support, so the correct start is from near the high. Traders new to this method will also run into trouble when they start at the top and then end the range at the capitulation spike low at point *1*. If you subdivide the range from the high to point *1*, all the Fibonacci subdivisions for 68.1 percent, 50 percent, and 38.2 percent will fall *above* the price low at point *2*. The error traders make is forgetting why they are making the calculation in the first place. Support levels must be below the current closing price. If the price range selected causes the Fibonacci subdivisions to fall above the current close, support has not been identified.

Point *3* is extremely common and a deadly error. Pay close attention to this one. Traders start the Fibonacci range at the high and then drag their mouse down to consider subsequent price lows. That is correct. However, *do not use any low that has been retraced by a more recent price swing.* If it has been

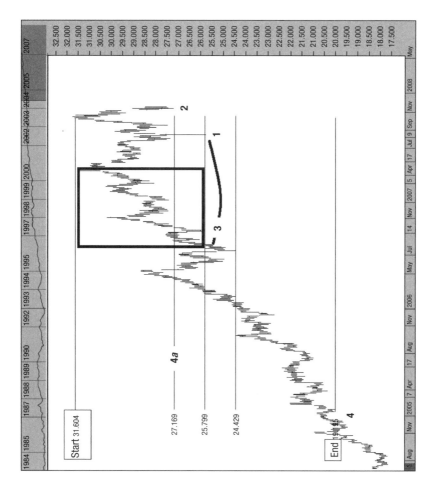

FIGURE B.1 Australia and New Zealand Banking Group Ltd. 3-Day Chart

Connie Brown, www.aeroinvest.com. Source: Copyright © 2008 Market Analyst Software

retraced, as we see in the decline from the high to point *1*, all the support levels behind and above it (those enclosed in the black box) are violated. So why use or consider them? Instead, scan down to point *1* and then immediately drag the cursor over to point *3*, and continue to a lower price low to end the range that will be subdivided. Keep in mind, confluence must occur when different Fibonacci ratios overlap, so you would continue to increase the size of the range subdivided by moving down to the next pivot, gap, or strong thrust price bar.

In **Figure B.2** we are looking at a 2-week chart for the Hong Kong WEB. The chart includes three subdivided ranges. A common error for a novice is to forget that multiple ranges will always start from the same price level. The top of the chart was truncated, and ranges that end at price lows *1*, *2*, and *3* were drawn. The confluence zone near 20.00 was respected by the market. This means the zones that form under this area are correct. When I truncate the start, often the lows that form spikes are not truncated. The market internals will tell you which is correct. However, do not pick ranges because of the internal alignments. The Nikkei 225 Stock Average will help you see why in the next example.

The chart in **Figure B.3** is a 3-day chart of the Nikkei. I want to find resistance in this example. My start for defining resistance must be a price low, and then I consider multiple price highs. In this chart there are two subdivided ranges. The first range ends at level *1* and the second at level *2*.

Notice the ranges and subdivisions *do not* extend to the left. I have not selected these ranges based on any data behind the current market swing down. Once people start to catch on to the concepts of confluence zones, they begin to pick ranges because they can make the market fit. Do not do this—you will fall into a huge trap. Here is why. The first confluence zone defining major resistance is at horizontal level *3* and price 14,269. I have extended the confluence zone slightly to the left in Figure B.3. If you looked at price lows marked *a* and *b*, you might think you made a mistake and then try to adjust the ranges. The next chart shows you why this would be a serious error to change the ranges to create what may appear to be a better confluence zone.

Figure B.4 shows the Nikkei 225 Index in a 3-week chart. You would normally never see the data extend back to 1992 in a single 3-day chart. But for this chart the x-axis was compressed to extend the historical view and the horizontal line created in the prior discussion

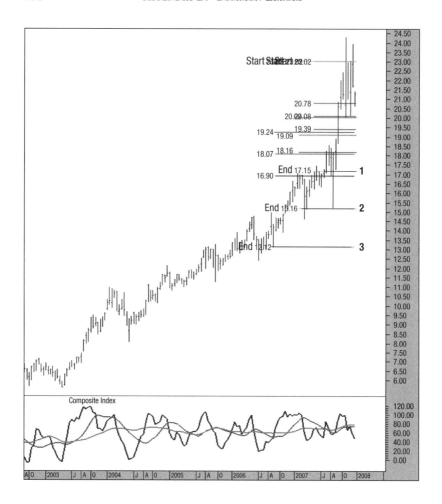

FIGURE B.2   **Hong Kong Index 2-Week Chart**

Connie Brown, www.aeroinvest.com. Source: Copyright © 2008 Market Analyst Software

was extended to the far left. It maps the exact lows at point *3* and *2* as support. Point *1* is the start of a major leg down failing at the same area as resistance. Keep in mind the horizontal line was the first confluence zone identified from short horizon data. Do not try to impose your will on the Fibonacci grids as they fall into place because you feel you have a better idea.

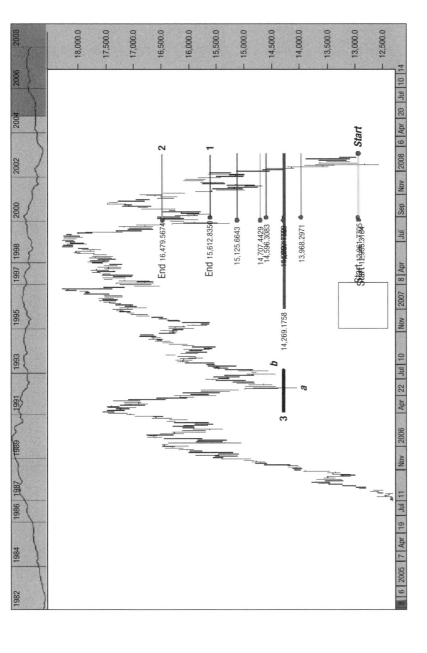

FIGURE B.3 **Nikkei 225 Stock Average 3-Day Chart**

Connie Brown, www.aeroinvest.com. Source: Copyright © 2008 Market Analyst Software

FIGURE B.4   Nikkei 225 Stock Average 3-Week Chart

Connie Brown, www.aeroinvest.com. Source: Copyright © 2008 Market Analyst Software

Selected Bibliography

Listed are the books, articles, and lectures consulted in the preparation of this book.

Ash   Ashton, Anthony. *Harmonograph: A Visual Guide to the Mathematics of Music.* New York: Walker & Company, 2003.

Bar   Barbera, C. André. "The Persistence of Pythagorean Mathematics in Ancient Musical Thought." PhD diss., University of North Carolina at Chapel Hill, 1980.

Bark   Barker, Andrew, ed. *Greek Musical Writings: II: Harmonic and Acoustic Theory.* Cambridge, UK: Cambridge University Press, 1989.

Benner   Benner, Samuel. *Benner's Prophecies of Future Ups and Downs in Prices.* 3rd ed. Cincinnati: Robert Clarke & Co., 1884.

Bru   Brunés, Tons. *The Secrets of Ancient Geometry—and Its Use.* Translated by Charles M. Napier. Copenhagen: Rhodos, 1967.

Bur   Burkert, Walter. *Lore and Science in Ancient Pythagoreanism.* Translated by E.L. Minar, Jr. Cambridge, MA: Harvard University Press, 1972.

Capt   Capt, E. Raymond. *A Study in Pyramidology,* Muskogee, OK: Artisan Publishers, 1986.

Case   Case, Paul Foster. *The True and Invisible Rosicrucian Order.* York Beach, ME: Samuel Weiser, 1989.

Chal   Chalmers, John H., Jr. *Divisions of the Tetrachord.* Lebanon, NH: Frog Peak Music, 1993.

Cog   Cogan, Robert. *New Images of Musical Sound*. Cambridge, MA: Harvard University Press, 1984.

Col   Colman, Samuel. *Nature's Harmonic Unity: A Treatise on Its Relation to Proportional Form*. Edited by C. Arthur Coan. New York: G.P. Putnam's Sons, 1912.

Cond   Condat, Jean-Bernard, ed. *Golden Section and Music*. New York: Peter Lang, 1988.

Cooke   Cooke, Roger. *The History of Mathematics*. New York: John Wiley & Sons, 1997.

Cooper   John M. Cooper, ed. *Plato: Complete Works*, by Plato. Indianapolis: Hackett Publishing Company, 1997.

Dew   Dewey, Edward R. *Cycles: The Mysterious Forces That Trigger Events*. New York: Hawthorn Books, 1971.

Doc   Doczi, Gyorgy. *The Power of Limits: Proportional Harmonics in Nature, Art, and Architecture*. Boston: Shambhala, 1994.

Dor   Dorfman, Allen Arthur. "A Theory of Form and Proportion in Music." PhD diss., University of California, Los Angeles, 1986.

Dunlap   Dunlap, Richard A. *The Golden Ratio and Fibonacci Numbers*. Singapore: World Scientific, 1997.

El-Daly   El-Daly, Okasha. *Egyptology: The Missing Millennium: Ancient Egypt in Medieval Arabic Writings*. London: UCL Press, 2005.

Far   Farmer, H.G. *The Sources of Arabian Music*. Leiden, Netherlands: E.J. Brill, 1965.

Fau   Fauvel, John, Raymond Flood, and Robin Wilson, eds. *Music and Mathematics: From Pythagoras to Fractals*. New York: Oxford University Press, 2003.

Fie   Field, Michael, and Martin Golubitsky. *Symmetry in Chaos: A Search for Pattern in Mathematics, Art and Nature*. Oxford: Oxford University Press, 1992.

Frey   Frey, Ronald Jan. "Attractiveness of the Golden Section in the Aesthetics of Sound Perception." MA thesis, University of Toronto, 1994.

Gad97   Gadalla, Moustafa. *Egyptian Cosmology: The Absolute Harmony*. Erie, PA: Bastet Publishing, 1997.

Gad99   ———. *Historical Deception: The Untold Story of Ancient Egypt*, Greensboro, NC: Tehuti Research Foundation, 1999.

Garl95   Garland, Trudi Hammel, and Charity Vaughan Kahn. *Math and Music: Harmonious Connections*. Palo Alto, CA: Dale Seymour Publications, 1995.

Gaz   Gazale, Midhat J. *Gnomon: From Pharaohs to Fractals*. Princeton, NJ: Princeton University Press, 1999.

Gies   Gies, Joseph and Frances. *Leonard of Pisa and the New Mathematics of the Middle Ages*. New York: Crowell, 1969; Gainesville, GA: New Classics Library, 1983.

Gillings Gillings, Richard J. *Mathematics in the Time of the Pharaohs*. Cambridge, MA: MIT Press, 1972; Mineola, NY: Dover Publications, 1982.

Godw   Godwin, Joscelyn, ed. *The Harmony of the Spheres: A Sourcebook of the Pythagorean Tradition in Music*. Rochester, VT: Inner Traditions International, 1993.

Hall   Hall, Manly P. *The Secret Teachings of All Ages: An Encyclopedic Outline of Masonic, Hermetic, Qabbalistic, and Rosicrucian Symbolical Philosophy*. New York: Penguin, 2003.

Hamb26   Hambidge, Jay. *The Elements of Dynamic Symmetry*. Originally published, n.p.; New York: Brentano's, 1926; New Haven, CT: Yale University Press, 1948; New York, Dover Publications, 1967.

Hamb32   ———. *Practical Applications of Dynamic Symmetry*. New Haven, Yale University Press, 1932.

Hark   Harkin, D. "On the Mathematical Works of François Edouard Anatole Lucas." *Enseignement mathématique* 3 (1957): 276–288.

Hea   Heath, Richard. *Sacred Number and the Origins of Civilization*. Rochester, VT: Inner Traditions, 2007.

Heath   Heath, Sir Thomas L. Introduction and commentary to *The Thirteen Books of Euclid's Elements*, by Euclid. 2nd ed. New York: Dover Publications, 1956.

Helm   Helmholtz, Herrmann von. *On the Sensations of Tone as a Physiological Basis for the Theory of Music*. New York: Dover Publications, 1954.

Herz   Herz-Fischler, Roger. *A Mathematical History of the Golden Number*. Originally published as *A Mathematical History of Division in Extreme and Mean Ratio*. Waterloo, Canada: Wilfrid

Laurier University Press, 1987. Reprint, Mineola, NY: Dover Publications, 1998.

Hof   Hofstadter, Douglas. *Gödel, Escher, Bach: An Eternal Golden Braid.* New York: Basic Books, 1979.

Kap   Kaplan, Aryeh (Rabbi). *Sefer Yetzirah: The Book of Creation: In Theory and Practice.* 2nd rev. ed. Newburyport, MA: Red Wheel/Weiser, 1997.

Kays   Kayser, Hans. *Textbook of Harmonics, Vols. I and II.* (Published originally as Lehrbuch Der Harmonik. Zurich: Occident-Verlag, 1950.) Edited by Joscelyn Godwin. Translated from the German by Ariel Godwin. Idyllwild, CA: Sacred Science Institute, 2006.

Kep   Kepes, Gyorgy, ed. *Module, Proportion, Symmetry, Rhythm.* New York: George Braziller, 1966.

Klei   Klein, Jacob. *Greek Mathematical Thought and the Origin of Algebra.* Cambridge, MA: MIT Press, 1968.

Kli   Kline, Morris. *Mathematical Thought from Ancient to Modern Times.* 3 vols. New York: Oxford University Press, 1972.

Kra   Kramer, Jonathan. "The Fibonacci Series in Twentieth-Century Music." *Journal of Music Theory* 17, no. 1 (spring 1973): 110–148.

Lach   Lachmann, R., and El-Hefni, M., translators. *Risala fi hubr ta'lif al-alhan (Über die Komposition der Melodien),* by al-Kindi. Leipzig, Germany: Fr. Kistner & C.F.W. Siegel, 1931.

Lar   Larson, Paul. "The Golden Section in the Earliest Notated Western Music." *Fibonacci Quarterly* 16, no. 6 (December 1978): 513–515.

Law   Lawlor, Robert, *Sacred Geometry: Philosophy and Practice.* London: Thames & Hudson, 1982.

Leh   Lehner, Mark. *The Complete Pyramids: Solving the Ancient Mysteries.* London: Thames & Hudson, 1997.

Lev   Levin, Flora. Translation and a commentary for *The Manual of Harmonics of Nicomachus the Pythagorean,* by Nicomachus of Gerasa. Grand Rapids, MI: Phanes Press, 1994.

Lew   Lewis, B., ed. *Islam and the Arab World.* New York: Alfred A. Knopf, 1976.

Liv   Livio, Mario. *The Golden Ratio: The Story of Phi, the World's Most Astonishing Number*. New York: Broadway Books, 2002.

MAC   Macran, H.S., ed. and trans. *The Harmonics of Aristoxenus*. Hildesheim; New York: Georg Olms Verlag, 1974.

Mad99   Madden, Charles. *Fractals in Music: Introductory Mathematics for Musical Analysis*. Salt Lake City, UT: High Art Press, 1999.

Mad05   ———. *Fib and Phi in Music: The Golden Proportion in Musical Form*. Salt Lake City, UT: High Art Press, 2005.

Mich   Michell, John F. *The Dimensions of Paradise: The Proportions and Symbolic Numbers of Ancient Cosmology*. Kempton, IL: Adventures Unlimited Press, 2001.

New   Newman, Rochelle, and Martha Boles. *Universal Patterns: The Golden Relationship: Art, Math & Nature*, Rev. ed. Bradford, MA: Pythagorean Press, 1992.

Neug   Neugebauer, O. *The Exact Sciences in Antiquity*. Originally published, n.p.; Providence, RI: Brown University Press, 1957. Reprint, New York: Dover Publications, 1969.

Parm   Parmanand, Singh. "Acharya Hemachandra and the (So Called) Fibonacci Numbers." *Mathematical Education* 20, no. 1 (1986): 28–30.

Pin   Pingree, David. "The Mesopotamian Origin of Early Indian Mathematical Astronomy." *Journal for the History of Astronomy* 4, no. 1 (1973): 1–12.

Rhind   Rhind Mathematical Papyrus. British Museum. Also see vol. 1. Oberlin, OH: Mathematical Association of America.

Robin   Robinson, Professor Daniel N. "The Great Ideas of Philosophy" lectures. Audio CD or DVD. Lectures from this Oxford University professor can be obtained from The Teaching Company, www.teach12.com.

Schoe   Schoenberg, Arnold. *Theory of Harmony*. Translated by Roy E. Carter. Berkeley: University of California Press, 1983.

Stan   Stanley, Thomas. *Pythagoras: His Life and Teachings*. (This is a photographic facsimile of the ninth section of the 1687 edition of Stanley's *History of Philosophy*.) Los Angeles, CA: The Philosophical Research Society, 1970.

Stroh   Strohmeier, John, and Peter Westbrook. *Divine Harmony: The Life and Teachings of Pythagoras*. Berkeley, CA: Berkeley Hills Books, 1999.

Taylor   Taylor, Thomas. *The Theoretic Arithmetic of the Pythagoreans*. Originally printed in London, 1816. Reprint. CA: Phoenix Press, 1934; York Beach, ME: Samuel Weiser, 1972.

Tom   Tompkins, Peter. *Secrets of the Great Pyramid*. New York: Harper & Row, 1971.

Vaj   Vajda, Steven. *Fibonacci and Lucas Numbers, and the Golden Section: Theory and Applications*. Chichester, England: Ellis Horwood, 1989.

Vor   Vorob'ev, N.N. *Fibonacci Numbers*. (Published originally in Russian as Chisla Fibonachchi. Moscow-Leningrad: Gostekhteoretizdat, 1951.) Translated by Halina Moss. New York: Blaisdell Publishing, 1961.

Index

About the Author

Connie Brown, CMT, founded Aerodynamic Investments Inc. (www.aeroinvest.com) after working for more than twenty years as an institutional trader in New York City. She continues to actively trade from her home in South Carolina and advises numerous financial institutions and banks around the world through the Internet. Many of her students have moved on to manage assets themselves or work for major institutions. She views seminars and lectures as an important part of contributing to the further development of technical analysis.

Brown's second book, *Technical Analysis for the Trading Professional*, was selected by the Market Technicians Association as required reading to prepare for the CMT Level 3 exam, the third and final examination that awards professionals the industry's Chartered Market Technician accreditation. Brown has written seven books and is the editor of the Market Technicians Association's *Journal of Technical Analysis*. She is a member of the American Association of Professional Technical Analysts.

About Bloomberg

Bloomberg L.P., founded in 1981, is a global information services, news, and media company. Headquartered in New York, Bloomberg has sales and news operations worldwide.

Serving customers on six continents, Bloomberg, through its wholly-owned subsidiary Bloomberg Finance L.P., holds a unique position within the financial services industry by providing an unparalleled range of features in a single package known as the Bloomberg Professional® service. By addressing the demand for investment performance and efficiency through an exceptional combination of information, analytic, electronic trading, and straight-through-processing tools, Bloomberg has built a worldwide customer base of corporations, issuers, financial intermediaries, and institutional investors.

Bloomberg News, founded in 1990, provides stories and columns on business, general news, politics, and sports to leading newspapers and magazines throughout the world. Bloomberg Television, a 24-hour business and financial news network, is produced and distributed globally in seven languages. Bloomberg Radio is an international radio network anchored by flagship station Bloomberg 1130 (WBBR-AM) in New York.

In addition to the Bloomberg Press line of books, Bloomberg publishes *Bloomberg Markets* magazine.

To learn more about Bloomberg, call a sales representative at:

London:	+44-20-7330-7500
New York:	+1-212-318-2000
Tokyo:	+81-3-3201-8900